THE GROWTH AND SUSTAINABILITY OF AGRICULTURE IN ASIA

by

Mingsarn Santikarn Kaosa-ard
and
Benjavan Rerkasem

with contributions by

Shelley Grasty
Apichart Kaosa-ard   Sunil S. Pednekar
Kanok Rerkasem   Paul Auger

OXFORD
UNIVERSITY PRESS

OXFORD

UNIVERSITY PRESS

Oxford University Press is a department of the University of Oxford.
It furthers the University's objective of excellence in research, scholarship,
and education by publishing worldwide in

Oxford New York

Athens Auckland Bangkok Bogotá Buenos Aires Calcutta
Cape Town Chennai Dar es Salaam Delhi Florence Hong Kong Istanbul
Karachi Kuala Lumpur Madrid Melbourne Mexico City Mumbai
Nairobi Paris São Paulo Singapore Taipei Tokyo Toronto Warsaw

with associated companies in Berlin Ibadan

Oxford is a registered trade mark of Oxford University Press

Published in the United States by Oxford University Press Inc., New York

Published for the Asian Development Bank by
Oxford University Press

British Library Cataloguing in Publication Data
available

Library of Congress Cataloging-in-Publication Data
available

ISBN 0 19 592450 9   (Paperback)
ISBN 0 19 592449 5   (Hardback)

Printed in Hong Kong
Published by Oxford University Press (China) Ltd
18th Floor, Warwick House East, Taikoo Place, 979 King's Road, Quarry Bay
Hong Kong

CONTENTS

FOREWORD

An economic transformation has occurred in much of rural Asia since the Asian Development Bank (ADB) last undertook a survey of the region in 1976. The rural economy has become increasingly linked to a rapidly integrating world economy and rural society in Asia faces new opportunities and challenges.

The transformation of rural Asia has also been accompanied by some troubling developments. While large parts of the region have prospered, Asia remains home to the majority of the world's poor. Growing inequalities and rising expectations in many parts of rural Asia have increased the urgency of tackling the problems of rural poverty. The rapid exploitation of natural resources is threatening the sustainability of the drive for higher productivity and incomes in some parts of rural Asia and is, in general, affecting the quality of life in the entire region.

These developments have altered the concept of rural development to encompass concerns that go well beyond improvements in growth, income, and output. The concerns include an assessment of changes in the quality of life, broadly defined to include improvements in health and nutrition, education, environmentally safe living conditions, and reduction in gender and income inequalities. At the same time, the policy environment has changed dramatically. Thus, there has arisen a need to identify ways in which governments, the development community at large, and the ADB in particular, can offer more effective financial and policy support for Asian rural development in the new century.

Therefore, the ADB decided to undertake a study to examine the achievements and prospects of rural Asia and to provide a vision for the future of agriculture and rural development in Asia into the next century. The objective of the Study was to identify, for the ADB's developing member countries in Asia, policy and investment priorities that will promote sustainable development and improve economic and social conditions in the rural sector.

The Study was designed as a team effort, using ADB Staff and international experts under the guidance of an ADB interdepartmental steering committee. To address the diverse issues satisfactorily and in a comprehensive manner, five thematic subject areas were identified to provide the analytical and empirical background on which the Study's recommendations would be based. Working groups comprising ADB staff were set up to define broadly the scope and coverage of each of the themes. The five working groups acted as counterparts to international experts recruited to prepare the background reports, providing guidance to the experts and reviewing their work to ensure high quality output.

A panel of external advisers from the international research community was constituted to review and comment on the approach and methodology of the study and the terms of reference for each of these background reports. The external advisers also reviewed the drafts of the reports. In addition, external reviewers, prominent members of academe and senior policymakers, were appointed to review each of the background reports and to provide expert guidance.

The preparation of the background reports included four workshops held at the ADB's headquarters in Manila: an inception workshop in May 1998; two interim workshops, in November 1998 and January 1999, respectively, to review progress; and a final workshop in March 1999, at which the background reports were presented by their authors to a large group of participants comprising senior policymakers from the ADB's developing member countries, international organizations, international and locally based nongovernment organizations, donor agencies, members of academe, and ADB staff.

The five background reports, of which this volume is one, have now been published by Oxford University Press. The titles and authors of the other volumes are:

Transforming the Rural Asian Economy: the Unfinished Revolution
Mark W. Rosegrant and Peter B. R. Hazell

Rural Financial Markets in Asia: Policies, Paradigms, and Performance
Richard L. Meyer and Geetha Nagarajan

The Quality of Life in Rural Asia
David Bloom, Patricia Craig, and Pia Malaney

The Evolving Roles of State, Private, and Local Actors in Rural Asia
Ammar Siamwalla with contributions by Alex Brillantes, Somsak Chunharas, Colin MacAndrews, Andrew MacIntyre, and Frederick Roche

The results and recommendations from the Study were presented at a seminar during the ADB's 32nd Annual Meeting in Manila. These have since been published by the ADB as a book titled *Rural Asia: Beyond the Green Revolution*.

The findings from the Study will provide a basis for future discussion between the ADB and its developing member countries on ways to eradicate poverty and improve the quality of life in rural Asia. The volumes in this series should prove useful to all those concerned with improving the economic and social conditions of rural populations in Asia through sustainable development.

Tadao Chino
TADAO CHINO
President
Asian Development Bank

PREFACE

This volume explores the transformation of agriculture in Asia since the last survey by the Asian Development Bank, published in 1978 when the impact of the green revolution was beginning to be felt and the focus of governments and researchers was on food security. In the last two decades, Asia has seen unprecedented growth in both the agricultural and nonagricultural sectors. Many countries in Asia have begun industrializing and their economies are no longer primarily agrarian in nature.

The volume is written within the rural Asian context and with special emphasis on sustainability issues. It is not intended to be a survey of Asian agriculture. Rather, it concerns how much growth there has been, what made that growth possible, and how growth can be further enhanced on a sustainable basis. Issues related to agriculture that are relatively remote in relation to rural Asia are given less attention. The focus of the book is how agriculture could become a component of a path to sustainable development.

The contributors to this volume are Mingsarn Santikarn Kaosa-ard, Benjavan Rerkasem, Apichart Kaosa-ard and Kanok Rerkasem from Chiangmai University; and Sunil Subanroa Pednekar, Shelley Grasty and Paul Auger from the Thailand Development Research Institute Foundation.

The authors wish to thank the following resource persons: Dr. Veravat Hongskul, Senior Fishery Officer, FAO Regional Office for Asia and the Pacific; Dr. Denis Hoffmann, Regional Animal Production and Health Officer, FAO; Mr. Henning Steinfield, FAO; and Mr. David Steane, Animal Genetic Resources Asia.

We would also like to thank the following who shared information, publications, and ideas with us: Drs. Peter Hazell and Mark Rosegrant, International Food Policy Research

Institute (IFPRI); Dr Larry Harrington, International Wheat and Maize Improvement Center (CIMMYT); Dr E.T. Craswell, International Board on Soils Research and Management (IBSRAM); Dr R. A. Fischer, Australian Centre for International Agricultural Research (ACIAR).

Finally, we gratefully acknowledge the assistance and support of the many Asian Development Bank personnel who were involved in the project, especially Bradford Philips and Shahid Zahid, and the members of the Working Group for this topic, in particular the Chair, Dimyati Nangju. The external advisers and external reviewers also provided useful information and suggestions on the content of the work.

MINGSARN KAOSA-ARD
and BENJAVEN RERKASEM

I THE PERFORMANCE OF AGRICULTURE IN ASIA

INTRODUCTION

Sustainability is a concept that has been gaining popularity since the 1980s. The most commonly cited definition of sustainability is that adopted by the Brundtland Commission: "development which meets the needs of the present without compromising the ability of future generations to meet their own needs" (WCED, 1987, p. 43).

For agriculture, the issue of sustainability is linked to that of food security, i.e. the sustained ability of agriculture to provide adequate food supplies. Concern about food security stems from the fear that as population increases, our ability to meet increasing food needs will be limited by the natural resource base. In addition, the technology of the green revolution, which was the introduction from the late 1960s of high-yielding varieties of rice, wheat and maize, application of chemical fertilizers and modern pest control methods, coupled with increased capital investment in irrigation and on farms, may have exhausted its potential. Furthermore, second-generation problems, which are related to the high-technology package and agricultural intensification, are claimed to be undermining future productivity through soil, water, and genetic degradation. Investment in irrigation infrastructure has also slowed down. Agriculture has encroached into wilderness lands, affecting biodiversity, which is fundamental to the sustainability of agriculture. The Food and Agriculture Organization of the United Nations (FAO) has estimated that between 1995 and 2010 the increase in agricultural cropland will place 85 million hectares (ha) of forests at risk. This trend of increasing threats to natural forests further exacerbates the possibility of climatic change through the release of additional carbon dioxide into the atmosphere. Indeed, the agricultural sector is being accused of undermining its own sustainability.

The concerns cited above are not at all recent. The issues concerning the possibility of sustained agricultural growth began in the 1940s and 1950s (Rattan, 1994), when the physical availability of natural resources was thought to be a possible limit to future growth. The second wave of concern, prevalent in the 1960s and 1970s, arose from the increasing intensification of agriculture and conflicts related to the multiple uses of natural resources and the environment, e.g. as inputs for production, recreational services, tourism sites, pollution sinks, and sources of potential future wealth (i.e. biodiversity). The third, and current, wave of concern was initiated by scientists in developed countries and deals more with global issues, such as climate change, ozone depletion, and acid rain.

In order to combat natural resource and environmental pressures, national and international research communities have joined forces in producing technologies that increase productivity, augment the existing natural resource endowment, and prevent food scarcity and starvation. The green-revolution technology was believed to be a win-win solution that overcame natural resource constraints and institutional changes.

The Brundtland Commission's definition of sustainable development, as cited above, applies to the concept in general. As far as agriculture is concerned, the Asian Development Bank (ADB) has another definition, namely "that which can evolve indefinitely toward greater productivity and human utility, enhance protection and conservation of the natural resource base, and ensure a favorable balance with the environment" (Tarumizu, 1992). This definition of sustainability is not just about maintaining environmental quality for a given level of resources. Nor is it about maintaining yields at current levels in perpetuity. The concept also includes (i) the need for enhancing productivity, and (ii) the need to meet increasing demands from growing populations. It is, therefore, not a static definition of constant production but refers to a *sustained increase* in production and consumption over time.

The ADB definition is particularly ambitious, considering that the standard economic interpretation of the Brundtland Commission's concept of sustainability requires that the per capita consumption of future generations remain at least as high as the current level. In order to maintain constant consumption levels over time, an amount equivalent to the economic depreciation of the exploited resources must be ploughed back into the investment as capital formation (Hartwick, 1977). This capital formation needed to replenish depleted stocks does not necessarily have to be physical capital. For agriculture, the ploughed-back amount could be in the form of investments in new technology and human resources. If consumption is allowed to increase over time, greater levels of plough-back investment are necessary. This volume adopts the ADB definition; the increase in yield levels or yield growth, a performance indicator for investment in technology, is used here as a proxy indicator for the need for more investment in technology to maintain agricultural sustainability.

This volume traces the past successes and the challenges yet to be overcome in achieving sustainable agriculture. The role of the State in management of technology transfer and of the natural resources sector is assessed vis à vis that of alternative institutions, such as the open market and local communities. We argue that technology, which has been a very powerful instrument in helping to meet food security needs in Asian countries, will not be able to continue in this role if policy and institutional reforms are not undertaken. This is especially true of those reforms related to natural resources and the environment. Current environmental degradation is a result of the mismanagement of technology, and failed policies and inappropriate government interventions. There are some early indications that the growth in productivity of rice production is leveling off, implying that the nature of research and technology development as well as the extension system will have to be modified.

We also argue that the yield gaps that continue to persist despite the green revolution are a reflection of the lack of attention that has been paid to less favorable environments. Past development efforts have concentrated on solutions designed in the laboratory rather than field-based crop management and, except for the People's Republic of China (PRC) and the transition economies, on technology and infrastructure rather than on policy and institutional reform. Although there is no large leap forward in productivity gains on the horizon for the next decade, substantial cumulative incremental gains could be made. The size of these gains depends on the ability of governments to fine-tune their research, development, and extension systems. This volume also emphasizes that agricultural sustainability can only come about if policies, including agricultural as well as economy-wide and natural resource policies, and institutions reflect environmental costs and demonstrate a proper understanding and appreciation of the complex relationships between nature, technology, and society.

THE SUCCESS AND SHORTCOMINGS OF THE GREEN REVOLUTION

The green revolution has been central to Asia's agricultural success. A key element is the use of new "improved" crop varieties developed with the aid of modern plant breeding techniques. Before the Second World War, Japan and its then colonies were the only Asian economies to employ crossbreeding extensively to increase crop productivity. Similar efforts did not begin in the rest of Asia until 1950, at which time breeding programs were instituted almost simultaneously in most Asian countries. International breeding programs began shortly afterwards, for rice in 1960 at the International Rice Research Institute (IRRI), for maize and wheat in 1966 at the International Center for Maize and Wheat Improvement (CIMMYT), and for soybean, mungbean, and some major vegetables at the Asian

Vegetable Research and Development Center (AVRDC). Modern varieties (MVs), the new varieties developed through both national and international breeding programs, began to be released and diffused in Asia beyond Japan, the Republic of Korea, and Taipei,China, from 1965.

The new varieties were generally superior in terms of yield potential, tolerance to pathogens and pests, and responsiveness to fertilizer and irrigation. They were also insensitive to photoperiod and/or required shorter growth time, making them more suitable for intensive cropping systems. The success of one group of MVs, the high-yielding varieties (HYVs), was internationally highlighted by the realization of a spectacular increase in output regionwide and the conferring of the Nobel Peace Prize on Norman Borlaug, who was the chief breeder of the technology development program at CIMMYT (Fairbain, 1995). The green revolution was the true Asian miracle of the 1970s and 1980s.

Without international assistance, the PRC was able to raise yield potential even further by developing, at the end of the 1970s, hybrid technology for rice and maize. The advent of the green revolution has saved Asia from famine and starvation (Box I.1). Nowhere has the impact of seed-fertilizer technology been greater than in South Asia where almost all countries have managed to feed their populations despite predictions of famine.

Bangladesh changed from being a net importer of 3.5 million metric tons (t) of grain annually in 1965 to self-sufficiency in grain by the early 1990s, by which time its population had grown from 53 million to 115 million (Gill, 1995). In India, where large-scale food shortages were avoided, the green revolution enabled food production to outpace population growth. Between 1970 and 1991, the annual rate of increase in food grain production was about 2.5 percent, while the annual rate of population growth over the same period was 2.2 percent. Technology had enhanced the food-growing capacity of India to the extent that it could have fed an additional 350 million people during that same period (Repetto, 1994). In the PRC, the food production index rose from 50 in 1975 to 145 in 1995, which implies that enough food was produced for an additional

Box I.1 Food and Famines

Globally, the world now produces enough food for its entire population. It is not the shortage of food but rather poverty, poor distribution, and mismanagement that have caused starvation and malnutrition. In 1943, inflation in Bengal drove food prices beyond the reach of the poor and caused 2-3 million deaths from starvation (FAO, 1995a). Major famines are mostly manmade, through war, ethnic or religious conflict, lack of foreign exchange, or abrupt economic crises, or are simply the result of inaccurate statistics and falsehoods.

The Great Famine that resulted in 30 million deaths in the PRC in 1959-1961 has been blamed on a number of factors. Explanations range from bad weather, inappropriate policies and incentives, poor reporting of crop yields, and even fraud. In order to satisfy the central leadership, local governments exaggerated grain output, leading to an excessive flow of grain out of the rural agricultural areas (Johnson, 1996). When the famine struck, the transportation system at that time and the sheer vastness of the country did not allow for the timely delivery of supplies to the deprived regions, resulting in one of the most devastating tragedies of our time. Some have claimed that it was forced collectivization that led to a decline in grain output in 1959 and 1960 (Chisholm and Jayasuriya, 1994). Lin (1988, cited in Lin, 1998a) suggested that it was due to "the deprivation of the peasant's right to withdraw from the collectives."

Brown (1995) warned that famine in the PRC may occur again. He estimated that by 2030 the PRC population will have increased by half a billion, putting tremendous pressure on the global food supply, and cited the increase in grain prices and large imports of grain by the PRC during 1994 as early signs of a growing imbalance between supply and demand.

In reality, the performance of agriculture in the PRC throughout the 1990s has been remarkable, except for areas affected by natural disasters. Cereal output rose steadily after 1950 through the middle 1990s. Wheat imports dropped from

(continued next page)

Box I.1 (continued)

7.2 million t in 1994 to 1.9 million t in 1997. In 1997, the PRC boasted a net export of about 1.1 million t of rice.

Highlighting famines and paying too much attention to statistics showing food production per capita may lead to "Malthusian optimism", i.e. the belief that raising the growth of food production per capita above the growth rate of the population will solve the starvation problem, which neglects the more pervasive and permanent problem of hunger and nutrition (Sen, 1986). Neither prices nor food production per capita are good warning signs or early indications of famines (Sen, 1986). More importantly, long-run food policies should not be limited to expanding food production per capita but should also enhance the ability of the individual to secure and be guaranteed food entitlements.

292 million people over that period. Similar success stories were repeated in Indonesia, Pakistan, Sri Lanka, and Viet Nam. Some countries, especially India, Thailand, and the Philippines, are now rapidly catching up with hybrid maize technology.

The green revolution not only helped to increase food production and supply, but also altered agricultural practices and cropping and trading patterns, and transformed rural livelihoods throughout Asia. The increased incomes and volume of trade encouraged associated activities such as food processing and transport. Expansion of electrically and mechanically powered irrigation and increased adoption of tractors and other farm machinery reduced the need for draft animals. A village-level study in Punjab, India, covering 1965 to 1978, revealed that camels were no longer used as draft animals and that the use of bullocks had decreased substantially (Leaf, 1984 and 1987, cited by Goldman and Smith, 1995). In their place, the numbers of food animals such as buffaloes and goats increased. Milk and meat became more readily available for household consumption. Rural poverty in India declined substantially as a result of government spending related to the green revolution (Fan, Hazell, and Thorat, 1998).

The impact of the green revolution on equity was questioned in early critiques. The technology involved can be seen to be selective and biased in favor of resource-rich regions and wealthy farmers. Fertilizer-responsive technology needs to be supported by a favorable environment, such as one with good irrigation, and tends to further aggravate the unequal distribution of income between resource-rich and resource-poor regions. Farmers also need credit worthiness, which tends to favor the large rather than the small farmer. Landless labor derives little benefit from these improvements, and employment levels have actually dropped due to the mechanization made possible by the higher productivity resulting from the green revolution. Rich and influential farmers were seen to maximize gains by ending tenancy agreements and lobbying for input and price subsidies (Fairbairn, 1995).

The increase in the supply of labor-intensive crops kept real wages low, which helped to support the expansion of labor-intensive enterprises. Fairbairn (1995) reviewed over 300 studies on the impact of the green revolution and found that 80 percent of the studies conducted between 1970 and 1989 concluded that the impact on equity of the green revolution was negative and that inequality increased during that period. It has been feared that the green revolution is a potential cause of increasing social antagonism and unrest (Frankel, 1971, cited in Sharma and Poleman, 1993).

The counter argument is that the negative effects on equity were the result of the early stages of the green revolution only. Citing field evidence from the northern Arcot region, Tamil Nadu, India, proponents of the green revolution indicated that the difference in yields between large and small farmers, evident in the 1970s, disappeared in the 1980s because smaller farmers were late adopters (Hazell and Ramaswamy, 1991). In fact, small rice farmers and the landless made larger gains in family income than did large rice farmers, farmers of other crops, and nonagricultural households. There was no increase in the concentration of land ownership. One study that found widened regional disparities in India between the mid-1960s and 1970s also found that a second-generation effect of the green revolution was increased output and profitability of small farmers. Other benefits have included widespread employment opportunities in postharvest operations such as storage, milling, marketing, and transportation (Sharma and Poleman, 1993). Increased rural incomes further brought about a diversification of rural economies and new opportunities for nonfarm activities. There was some loss in employment because of mechanization and the use of pumping for irrigation, but improvements in real wages led to increased earnings for the landless and nonagricultural households.

Another study (David and Otsuka, 1994) on the impact of adoption of HYVs in seven Asian countries (Bangladesh, PRC, India, Indonesia, Nepal, the Philippines, and Thailand) concluded that although HYVs improved productivity in favorable (irrigated) areas relative to that in less favorable areas, other indirect effects have tended to prevent significant worsening of disparities in income distribution. These indirect effects have included increased real wages in unfavorable areas through migration out to favorable areas where employment opportunities are higher, decline in the real price of rice, which has benefited consumers, and changes in land tenure that have mitigated the worsening of disparities in income distribution. An exception is the development of hybrid rice in the PRC where there has been a direct positive impact on equity, because the new rice was adopted in the mountains in unfavorable regions.

A more recent study (Hazell and Fan, 1998), on marginal returns to technology inputs in India in 1994, found that the marginal return in rainfed areas from government HYV expenditure was almost twice that in irrigated areas. On the basis of State-level data for 1970 to 1994, the authors confirmed that increased agricultural productivity reduced poverty directly by increasing farmer income and indirectly through employee wages and reduced agricultural prices. Poverty of the landless increased, although to a small extent.

Most studies on the impact of income distribution concentrate on income from rice farming. When the total income of all households, both those adopting and those not adopting the new technology, is considered, the impact of income distribution on rural households is negligible (Lin, 1998b). In his study of 500 households in Hunan Province, Lin found that technology adopters tended to increase the amount of resources allocated to rice production relative to other activities, while the reverse was true for those not adopting the new technology. Therefore, if rice is the only source of income considered, an inequality is to be expected because the nonadopters tend to reduce the amount of resources committed to rice production. By examining total income for all outputs, the impact on equity of the new technology is seen to be minimal.

Later critiques of the green revolution have focused on its ecological and biological impact. The high-technology package used has disturbed the ecological equilibrium, creating undue dependence on external inputs and stretching the Earth's support system beyond its capacity. The spread of HYVs, which have a narrow genetic base, increases the risk of greater exposure to pest and insect attacks. The associated intensive use of agrochemicals could have a negative impact on the quality of water, harming the health of farmers, consumers, farm animals, wildlife, and the environment. Also, since the green-revolution technology concentrates on a few staple crops grown in favorable regions, farmers in unfavorable areas have no option but to engage in extensive agriculture, resulting in encroachment into natural forests and fragile ecosystems. The green-revolution package has inherent weaknesses and second-generation effects associated with its high input practices. Finally, the technology involved depends on fossil-fuel energy sources, which are nonrenewable. This could undermine the long-term sustainability of green-revolution technology.

The above arguments are examined in later sections. Here, it is sufficient to note that the least recognized, but probably greatest, benefit of the green revolution is that the increase in food output has reduced the need for opening up more land for agriculture, especially in the more fragile ecosystems. This has prevented large-scale deforestation. It is estimated that without the green revolution, at least 60 percent more land would be required to maintain the current population at the prevailing nutritional standards (ODI, 1994, cited in Gill, 1995). In 1985, The Consultative Group on International Agricultural Research (CGIAR) estimated that without the modern varieties about 20–40 million ha more would be needed to produce rice and maize in the humid tropics (CGIAR, 1985, cited in Harrington, 1993). The various criticisms should not be taken as a rejection of the green revolution or of the value of an increase in food supply and food security. Rather they should be taken as providing directions for future research and improvement.

AGRICULTURAL GROWTH TRENDS (1967-1997)

The ability to meet increasing demands from growing populations requires that production growth exceeds population growth (Table I.1). The green revolution has made this possible over the last few decades. Demand and supply projections up to 2010 indicate that production growth of cereals will be high enough to allow a slight fall in the real price of food. Rice is the only major staple crop for which prices may increase (Rosegrant and Hazell, 1999). Some other developments occurring in tandem with the green revolution have been innovations in the areas of livestock, aquaculture, and coastal and oceanic resources. This section examines the growth trends and environmental impact of the food sectors as well as those of tree plantations, the latter being brought into the analysis for their relatively more benign impact on the environment and their implications for land use.

Annual Crops

Asia contributes over 90 percent of the world's production of rice, about one third of all wheat and about one fifth of all coarse grain (Khan, 1996). Three major trends can be observed in the yields of field crops in Asia. First, growth in the production levels of food crops, mainly cereals and pulses, has been decelerating during the third decade of the green revolution (Table I.1). The yield increases of cereals and pulses peaked at 3.8 percent per annum during 1977–1986, but slowed to 2.3 percent during the next decade. However, the latter growth rate was still above that of population growth for the decade. The average annual yield growth of crops other than cereals and pulses rose from almost zero from 1977 to 1986 to almost 2 percent in the following decade. Asia's population grew at an annual rate of 1.82 percent during 1987–1997, down from 1.87 percent a decade earlier (Annex Table A1).

Table1.1: Crop Production in Asia, 1977–1997

| | Average Growth (percent per year) | | | |
| | Production | | Yield | |
	1977–1986	1987–1997	1977–1986	1987–1997
Cereals and pulses	3.82	2.60	3.80	2.29
Others[a]	3.22	5.16	0.20	1.81
Total	3.47	4.08	2.51	2.71

[a] includes fibers, oils, roots, sugar, tea, coffee, tobacco, rubber, vegetables, fruits, and nuts.

Source: FAOSTAT Database. *Available: http://apps.fao.org*

The second trend is a shift away from, or a fairly strong diversification out of, food grains in favor of higher value crops (Table I.2, Annex Tables A2, A3, and A4). The decline has been most drastic for rice, millet, and sorghum (Annex Table A5). The trend has also reduced the dominance of food grains in the total cropping system. In the PRC, the loss of land sown with food grains was substantial, amounting to 8 million ha or about 10 percent of total harvested food grain area. The reasons for the shift to nonfood grains were the decline in real prices of food grains (Beckerman, 1995), a declining profitability due to a price/cost squeeze, and an increased demand for high-value horticultural crops.

Table 1.2: Diversification of Asia's Cropping System
1977–1997

	Area				Average Growth	
	ha, million		percent of total		(percent per year)	
	1976–1978[a]	1995–1997	1977	1996	1977–1986	1987–1997
Cereals and pulses	304.570	307.554	73.84	63.60	0.03	0.25
Others[b]	107.919	176.034	26.16	36.40	3.06	2.78
Total	412.489	483.588	100.00	100.00	0.82	1.05

[a] three-year mean.
[b] includes fibers, oils, roots, sugar, tea, coffee, tobacco, rubber, vegetables, fruits, and nuts.
Source: FAOSTAT Database. Available: http://apps.fao.org

Third, despite criticism voiced since the 1970s that the seed-fertilizer package is beneficial only in favorable environments, little progress on technology applicable to less favorable areas has been made. For example, rainfed rice yields are only half of those of irrigated agriculture, with even lower yields for upland and deepwater areas (Rosegrant and Pingali, 1990). Maize yields in Rajasthan, Uttar Pradesh, and Madya Pradesh in India and in many other countries are low. The exceptions are countries where rapid diffusion of hybrid maize has occurred, such as the PRC, Thailand, and the Philippines.

Amongst the major cereals, wheat and maize continued to show robust yield growth during the last decade (Annex Table A2). Rice is the only staple for which the yield growth, or the average annual increase in output per hectare, has fallen below 2 percent, almost half that of the preceding decade.

The highest average annual rice yields in Asia were 6.564 and 6.545 t/ha in 1977 and 1997, respectively, both in the Republic of Korea (Annex Table A6). In the PRC, yields jumped to 6.187 t/ha from 3.704 t/ha. For Asia as a whole, yields went up from 2.596 to 3.840 t/ha during the same period. The Asia-wide average, annual yield growth rate, however, did decline substantially, from 3.35 percent during 1977–1986 to 1.50 percent during 1977–1996. This deceleration reflects the fact that the growth potential of the early innovations of the green revolution has been exhausted in the best-suited areas. IR8, developed by IRRI, was the first HYV to break the yield barrier for rice and started the green revolution outside the PRC. In fact, other HYVs have never been able to better the yields of IR8, although the progress made in pathogen and pest resistance in later HYVs has been remarkable. Considering that 74 percent of wet riceland is sown with MVs, there is now a renewed need for new rice varieties that will raise the yield potential.

Examining the rice industry more carefully also reveals that the slow down in yield growth could be related to the diversification of MVs into traditional varieties that have higher eating quality but lower yields. For example, yield declines in Karnal, Haryana, in the heart of India's green revolution riceland, were a result of the move to grow *basmati* rice (Chaudhary and Harrington, 1993). A similar but stronger trend has been observed along Asia's Pacific rim where a boom in manufacturing has increased the opportunity cost of agricultural labor. While the high-quality rice production area has expanded in response to greater demand and better prices, the production of low-quality rice and also of marginal food crops has declined. However, migration from rice to more lucrative crops and to higher paying jobs in the city has also increased the cost of rice.

Rapid economic growth in the cities has fuelled the demand for high-quality rice and horticultural products. In the southern PRC, the heartland of rice production in that country, a drastic decrease in the area being sown has been observed (Hong, 1996). In Thailand, where the industrial boom continued unabated for a decade before coming to an abrupt end in 1997, productive resources, especially labor, moved away from the agricultural sector in general and the rice sector in particular as a result of competition for these resources (Ammar, 1996; Coxhead and Jiraporn, 1998). Rapid growth in other newly industrialized economies has also placed pressure on rice production in terms of labor and land costs. Economic recessions will dampen demand somewhat, but this does not reduce the need for an effort to increase rice productivity.

In contrast with rice, the yield growth in wheat and maize remains strong (Annex Tables A7 and A8). This is particularly true of wheat in the PRC, India, and Pakistan. For maize, growth potential can be further tapped in favorable areas in Cambodia, India, Indonesia, the Philippines, and Viet Nam.

The relatively strong growth trends in wheat are partly related to the success of breeding programs, which have raised the yield potential of new varieties at a rate of about 1 percent per year, while maintaining their resistance to the main pathogens. For example, the average number of newly released wheat varieties per year in India increased from 2.6 in the 1960s to 3.4 in 1970s and 7.2 during 1980 to 1985 (Byerlee, 1990). The yield potential of wheat MVs in South Asia has increased at a rate of 0.5 to 1 percent per year due to genetic improvement. The historical trend of productivity gains in the PRC has continued, with annual productivity gains averaging 3.33 percent from 1987 to 1997 (Annex Table A7), a further suggestion that the sustainability of the yield potential for wheat should not be a great cause for concern. However, as with rice the profitability of wheat is decreasing, for example in areas like Karnal (Chaudhary and Harrington, 1993). This should be a greater cause for concern.

The rapid growth in maize production is a response to the boom in the poultry industry, which is a major consumer of maize. The PRC has produced three generations of hybrid maize in the last three decades with a 10 percent productivity gain from each new generation (Chen, 1995). Elsewhere, hybrid maize was adopted much later, spearheaded mainly by the seed or feed businesses in the private sector. Recent successes in Southeast Asia, starting with open-pollinated varieties followed by hybrids, imply that the potential exists for an expansion of maize output and increase in yield potential in the tropical environment.

Asia dominates the world in both rice production and rice consumption, accounting for over 90 percent of each (Hossain, 1996). Rice is consumed universally in Asia, whereas wheat is consumed as a major part of the diet only in Pakistan and some parts of Bangladesh, PRC, India, Nepal, and Central Asia. Wheat output in Asia is about 40 percent that of rice, and land sown with wheat covers about 60–70 million ha, about half the rice-growing area, 130 million ha (Annex Table A2). The amount of land sown with maize (35 million ha) is only about one third of that sown with rice. This relative order of magnitude is important when planning the allocation of R&D expenditure on food crops in Asia.

Despite the fact that rice production in Asia is enormous, the internationally traded volume is small, accounting for only a small proportion of the total production–3 percent during 1994 to 1996–while 18 percent of wheat was traded internationally in the same period. The small international rice market implies that a reduction in production by any major rice-consuming country will have a significant impact on international trade and that rice prices may potentially vary considerably. This in turn implies that the poor in Asia are relatively more vulnerable with respect to food prices. The low levels of foreign exchange held by some Asian governments following the financial and economic crisis in Asia may further aggravate food security.

Sorghum, millet, and barley have shown a declining trend in production and area sown over the past 20 years, although yield growths have been positive and increasing at between 1.4 and 1.8 percent per year (Annex Table A2). Among other major crops, there has been an increase in oilseed production, particularly for soybean, rapeseed, sunflower, and castor during this period, while tubers have shown different growth trends. Cassava production declined from an average annual increase of 2.9 percent for 1977–1986 to negative 0.68 percent for 1987–1997, although potato production has shown signs of remarkable growth over these two decades.

Crop production is thought to affect the environment not only directly, through nutrient depletion and the emission of greenhouse gases, but also indirectly where expansion of crop areas is a threat to forest areas. These issues are discussed further in later chapters.

Perennial Crops

The major perennial crops in Asia, as measured in terms of harvested area (above 1 million ha) in 1997, are coconuts (9.1 million ha), sugar cane (8.6 million ha), rubber (6.7 million ha), oil palm (4.7 million ha), tea (2 million ha), and coffee (1.6 million ha). Together, the harvested area of coconuts and sugar cane (17.7 million ha) is less than half of the area devoted to maize and close in size to the area devoted to fruits (16.8 million ha) and vegetables (20.8 million ha). In Asia, perennial crops may be grown on large plantations, in smallholdings, or under a subcontracting system as is the case with the Thai sugar cane industry.

Tree-based systems, when properly managed either as plantations or agroforestry systems, have a relatively benign impact on the environment, especially when grown in the uplands and on steeper slopes. Also, once an investment is made in trees, the land is committed to a fixed pattern and therefore cannot be easily converted into land for food crops, at least in the short term.

The largest producers of coconuts in 1997 were Indonesia (14.7 million t), the Philippines (12.1 million t), India (9.8 million t), and Sri Lanka (2 million t) (Annex Tables A9, A10, and A11). In the last decade, a small decline in harvest areas has occurred only in the Philippines. The World Bank has forecast a substantial decline in prices for both copra and coconut oil until 2010.

Southeast Asia produces almost all of the world's rubber output of 5.2 million t (Annex Tables A12, A13, and A14). Thailand is the largest producer, contributing on average about 2 million t of output, which is about one third of the world's total natural rubber production, from harvested areas that covered 1.5 million ha in 1997. The second largest producer, Indonesia, produces on average 1.5 million t per year from 2.3 million ha of plantations, followed by Malaysia, with about 1 million t from 1.5 million ha. These three countries together produce about three quarters of the world's total output. Harvested areas in Malaysia are experiencing a declining output, while Indonesia has seen a slight increase over the last seven years. Countries showing strong increasing trends but from relatively small bases are Viet Nam, Myanmar, and Cambodia. The World Bank's forecast for rubber prices indicates a short-term price drop followed by a recovery towards 2010.

Tea is a traditional crop in many Asian countries, but it is grown on a relatively large scale only in the PRC (0.9 million ha), India (0.4 million ha), Sri Lanka (0.19 million ha), and Indonesia (0.11 million ha) (Annex Tables A15, A16, and A17). The total area used for tea production has been relatively stagnant since 1975. India and Indonesia are the only countries showing slight increases in harvest areas. Sri Lanka, a world-renowned producer of black tea, has experienced a clearly decreasing trend in area under tea.

Harvested areas for coffee in Asia total 1.6 million ha and almost all of this is in Southeast Asia, i.e. Indonesia (0.8 million ha), India (0.24 million ha), Viet Nam (0.19 million ha), and the Philippines (0.15 million ha). The area under coffee is increasing strongly in Indonesia and Viet Nam and slightly in India (Annex Tables A18, A19, and A20).

Oil palm is not yet a very widespread crop in Asia, but the area under harvest is growing rapidly in Indonesia, Thailand, and Malaysia (Annex Tables A21, A22, and A23). The crop has high yield potential and requires a relatively small amount of labor for planting, maintenance, and harvesting. It demands a warm climate and evenly distributed rainfall, making Indonesia, Malaysia, and southern Thailand suitable growing areas. In 1997, the industry suffered from the widespread fires in Sumatra and Kalimantan, Indonesia (Box I.2). It can be expected that further expansion of oil palm will be somewhat hindered by the financial and economic crises occurring in the three major producing countries. Moreover, the World Bank has forecast a rapid decrease in the price of palm oil until the year 2010 due to excess competition and production.

Land clearing for agricultural tree plantations in Sumatra was proven to be partly responsible for the fires in 1997 that erupted into a regional environmental problem. The fires, exacerbated by the long drought associated with El-Niño, lasted from mid-1997 to early 1998, producing enormous quantities

Box I.2 The Great Haze: Who was Responsible?

Fire has always been a part of agricultural management in Asia because it is the cheapest and least capital-intensive method of clearing land (ICRAF, 1996). It reduces pests, diseases, and weeds. Farmers also believe that burning increases soil fertility. Fire is used as a land-clearing tool by both smallholders and large plantation companies. Traditionally, institutions in communities engaged in slash-and-burn agriculture monitored and enforced measures, such as fines and other penalties, to ensure that fires did not go out of control. However, local communities have no control over the large companies that also use fire for clearing land to establish plantations.

Before the great haze of 1997, smallholders had often been blamed as the cause of forest fires in Indonesia. An advanced, high-resolution oceanic and atmospheric satellite identified 12,000 fire spots in Sumatra in September and October 1997. The fires were not really 'forest fires', as only 44 percent of the hotspots were in forest areas. Satellite images verified that the large companies were in fact responsible for the fires.

Researchers have indicated that under the existing development policies, fire is not unexpected and will return with or without El Niño (Tomich et al., 1998b). The Government provides incentives and grants land to large companies for the development of large-scale plantations, but does not recognize the rights of the local farmers occupying the lands provided to companies for plantations. Fire has thus become a 'weapon' of the companies to get rid of smallholders, and vice versa. Moreover, Indonesia's policies tend to favor export of sawn timber rather than roundwood or logs in order to nurture forward linkages, which encourages the treatment of wood felled during clearing as waste, and discourages the protection of standing timber. In addition, peat swamps are converted for rice production. Fires from peat forests tend to linger underground and are difficult to extinguish.

Alternatives to slash-and-burn agriculture are now being sought for small farmers by the International Center for Research in Agroforestry (ICRAF), but for the time being, banning the use of fire by smallholders is impractical. Attempts must be made

(continued next page)

Box I.2 (continued)

to solve the problem in a holistic manner. A number of policies have been suggested, such as a ban on rice production in peat swamps, finding land for plantation agriculture in grasslands rather than forests, recognition of land rights of local communities, the revision of promotion incentives and conditions for large-scale plantations, and a review of forest-product export policies. At the regional level, a system with shared responsibilities needs to be devised to effect better management and monitoring and to improve fire-fighting capacity.

of smoke and haze that covered Indonesia, Singapore, Malaysia, and southern Thailand. In Indonesia, 5 million ha of forest and agricultural lands were damaged and 70 million people in the region were affected (EEPSEA and WWF, 1998). The haze also seriously affected the previously booming tourism industry in the four countries. Total damages (calculated up to December 1997) to Indonesia were estimated at $3.8 billion[1] while about $670 million worth of damage was done to neighboring countries. Development policies favoring the conversion of forest into plantation have been seen as a major contributor to the haze problem (Murdiyarso, 1998).

Well-managed tree plantations tend to be less harmful to the environment than some field and garden crops. Well-managed plantations with good ground cover can reduce the rate of run-off and erosion to below 5 t/ha per year, which is a better rate than that of degraded forests and shrubs. For example, a well-managed tea plantation results in an annual loss of only 0.24 t/ha of soil, compared with 25–100 t/ha for vegetables, potatoes, and tobacco, and 0.3 t/ha for dense forest (Chisholm, Ekanyake, and Jayasuriya, 1997). In the early stages

---

[1] $ indicates US dollars throughout the text.

of establishment, erosion rates tend to be high but the mulching done during the first two years of planting can considerably reduce run-off and soil erosion.

Asia is a big producer and consumer of sugar. It has five of the world's top ten consumers, namely India, PRC, Indonesia, Pakistan, and Japan. The top five producers of sugar in 1997 were India (277 million t), PRC (82.57 million t), Thailand (45.85 million t), Pakistan (42 million t), and Indonesia (27.76 million t) (Annex Tables A24, A25 and A26). The Philippines was a much bigger producer than both Pakistan and Thailand in the 1970s and early 1980s but the latter two countries increased their capacity substantially during the 1990s. Asia produces about one third and consumes about 45 percent of the world's sugar output. Hence, the region has a sugar deficit. The annual rate of growth of sugar cane production has also declined, from 4.9 percent in the 1950s to 1.6 percent in the 1990s.

The sugar cane industry in Asia is dominated by smallholders, implying a need for an efficient institutional arrangement between them and factories. Yields are highest for Indonesia and the PRC (about 71 and 75 t/ha, respectively), both of which benefit from irrigation (Annex Table A26). Yields in the Philippines and Thailand, which use relatively low-input rainfed systems, are about 69 and 49 t/ha, respectively. However, Thailand has the highest average sucrose content at 13 percent, followed by India (12 percent), Philippines (11 percent), and Indonesia (9.6 percent). Thailand has the lowest production cost, followed by India and Indonesia (Fry, 1998).

As a C4 plant[2], sugar cane has relatively efficient photosynthesis and hence absorbs more CO_2. However, the

---

[2] C4 plants have a special CO_2 -concentrating mechanism within their leaves by which they can increase the CO_2 concentration to several times that of ambient levels.This is done by CO_2 first being incorporated into a 4-carbon compound. This allows these plants to maintain lower intercellular CO_2 concentrations than C3 plants. C4 plants tend to grow in warmer, more water-limited regions, and include many tropical grasses and the agriculturally important species maize, sugar cane, and sorghum (IPCC, 1996).

current harvesting method in Asia involves burning, which releases the absorbed CO_2. Green harvesting methods have emerged and need to be introduced in Asia. This will be more important to Asian exporters when the EU market, which is now highly protected, opens up.

Forest Plantations and Agroforestry

Forest plantations in the region, such as teak plantations in India, Myanmar, and Thailand, were established early in the 20th century. These plantations were established mainly by government organizations for several reasons, including production and conservation. The harvesting rotations vary from medium, i.e., 20–30 years, to long, 40–80 years, aiming primarily to produce high-quality timber for the international market. Currently, most of the plantations are State owned. Since 1980, the private sector has become increasingly involved in setting up large-scale plantations of species that are especially fast growing, such as *Acacia, Albizzia, Eucalyptus, Gmelina, Paraserianthes*, and bamboo in the PRC, India, Indonesia, Malaysia (Sabah and Sarawak), the Philippines, and Thailand. The primary objective of these commercial or private plantations is to produce industrial raw material, including roundwood and pulpwood, with short and medium harvesting rotations of 10–20 years. This objective is being promoted by the governments concerned. With very few exceptions, modern breeding techniques and improved clones are increasingly used in Asian plantations. There are also a few examples of manmade forests, established by indigenous people using traditional knowledge (Box I.3), and which have gained national and international recognition.

Forest plantations in Asia in 1995 totaled 59 million ha or about 15 percent of the area of natural forests. Between 1980 and 1995, Asia lost 63.26 million ha of natural forest cover (FAO, 1995b, 1997a). The total area devoted to forest plantations is about nine times that occupied by rubber plantations. Despite the fact that about four fifths of the existing plantation areas

Box I.3 The Manmade Forest of Krui, Lampung Barat, Indonesia

Privately operated plantations are generally large and are established using capital-intensive technology. They are usually planted with introduced species. In the village of Pahmungun, Lampung Barat, a damar *(Shorea javanica)* forest was established by the indigenous communities in the 1890s. The damar tree was domesticated along with coffee and fruit trees. It is a dipterocarp resin-yielding tree and is valued both for resin and timber. Damar agroforest owners are estimated to earn slightly greater incomes than owners of monoclonal rubber plantations. Apart from supplying 70 percent of the annual cash income of villagers, the damar forest also provides ecological functions and acts as a storehouse of biodiversity. The forest, approximately 55,000 ha, supports 17 species of rare plants, 17 species of protected mammals, and 92 species of birds.

Until the discovery and identification of the human contribution to its sustainability, the damar forest of Krui was simply assumed to be a natural forest. In 1998, the Indonesian Government issued a decree that acknowledged and legitimized the indigenous land-use system in State forestland as a distinct forest-use classification. Under this decree, local people are allowed to harvest the timber they have planted in State forests. A limited amount of logging is also allowed in the watershed forests. Local communities are given the right to manage a part of the State forests under their traditional customs. The Krui forests were among the first to be provided with such rights.

Sources: de Foresta and Michon (1997); Fay et al. (1998).

are used for nonindustrial purposes, e.g. for conservation and for household consumption or community uses, forest plantation establishment rates have lagged far behind deforestation rates.

The PRC has the largest plantation area, with an annual planting rate of 1.1 million ha (Table I.3). Most of the PRC's plantations are aimed at conservation and nonindustrial purposes. Recent devastating flooding has prompted the Government of the PRC to stop commercial logging in the western watersheds.

India has the largest area of industrial plantations, totaling 5.7 million ha, consisting of fast-growing species (5-10 year rotation) covering 0.9 million ha, and other industrial species covering 4.8 million ha. In Southeast Asia, Indonesia has the largest area of forest plantations, 6.1 million ha, one third of which is planted for industrial purposes. Lao PDR lies at the

Table I.3: Forest Plantations by Type in
Selected Countries in Asia in 1990
(annual change of planting area during 1981–1990)

	Plantation Area (ha'000)				Annual Change (ha)
	Fast Growing	Other Industrial	Non-industrial	Total	
China, People's Rep. of	2,120	1,000	28,711	31,831	1,140
Taipei,China	0	0	10	10	
East Asia, total	2,120	1,000	28,721	31,841	1,140
Cambodia	0	0	7	7	
Lao PDR	0	3	1	4	0.1
Myanmar	0	155	80	235	19.6
Thailand	180	85	264	529	29.4
Viet Nam	560	0	910	1,470	49.0
Indonesia	1,150	280	4,695	6,125	331.8
Malaysia	80	0	1	81	6.3
Philippines	1	5	143	149	
Southeast Asia, total	1,971	528	6,101	8,600	436.2
Afghanistan	0	0	8	8	
Bangladesh	50	85	100	235	12.3
Bhutan	0	4	0	4	0.2
India	900	4,770	7,560	13,230	1,009.0
Nepal	10	10	36	56	4.3
Pakistan	0	50	118	168	4.2
Sri Lanka	30	95	14	139	6.0
South Asia, total	990	5,014	7,836	13,840	1,036
Total	5,081	6,542	42,658	54,281	2,612.2

Sources : ADB (1995a); FAO (1997a, 1997b).

other extreme, having the smallest plantation area, 4,000 ha, most of which is recent and geared towards the pulp industry. Among developing countries, the proportion of industrial to nonindustrial plantation area is highest in Lao PDR, Myanmar, Malaysia, and Sri Lanka.

Although there are many different tree species planted, especially in tropical Asia and the Pacific countries, eucalyptus appears to be the preferred species group of regional planting programs. This species group accounts for about 16 percent (5.2 million ha) of the total plantation area in the region. Acacia, teak, and pine are also major groups, accounting for 11 percent (3.4 million ha), 6 percent (2.19 million ha), and 4 percent (1.25 million ha) of total plantation area, respectively (FAO, 1995b). The remaining 64 percent (20.6 million ha) of plantation area contains other or unclassified tree species such as *Albizzia*, *Dalbergia, Casuarina, Leucaena, Swietenia, Xylia, Gmelina*, and *Pterocarpus*.

Eucalyptus and acacia plantations are established primarily for the pulpwood and medium-fiber wood industries. They are also favorites of nonindustrial plantations established for community use, e.g. for fuelwood, small wooden poles, for land rehabilitation, and for environmental conservation. The eucalyptus species planted in the region include *Eucalyptus camaldulensis, E. deglupta, E. europhylla, E. globulus*, and *E. grandis*, the hybrid species. The most common acacias are *Acacia auriculiformis, A. mangium, and A. nilotica*, and the pines most often found are *Pinus kesiya, P. merkusii, P. caribaea*, and *P. oocarpa*.

The growth and yields of these species vary according to onsite characteristics and genetic material (improved seed sources and clones). In Indian and Indonesian teak plantations, the yield of an average plantation is about 2–3 m^3/ha/year after about 50–70 years. The average yield of eucalyptus plantations is about 6 m^3/ha/year after 8–10 years. The average yield of *Acacia mangium* plantations in Malaysia and Indonesia is more than 20 m^3/ha/year after 5 years.

Scientific progress relating to the establishment of forest plantations and logging operations has been stagnant because government agencies or State enterprises run these activities. Unlike forest plantations in developed countries or those operated by multinational corporations, forest plantations in Asia are run under low-input, low-output systems. Exceptions are found in the large-scale plantations in Sabah.

The impact of forest plantations on soil erosion is indicated in Table I.4. Surface erosion from well-managed forest plantations is small, averaging around 0.6 t/ha/year. However, if forest litter is removed for use as fuel, as is the case in the PRC, erosion can be much greater than for shifting cultivation.

Table I.4: Surface Erosion in Tropical Forests and Tree Plantations

Forest and Tree Plantations	Annual Soil Loss (t/ha)	
	Range	Mean
1. Natural forests	0.03–6.20	0.3
2. Shifting cultivation during fallow-period years	0.05–7.40	0.2
3. Forest plantations	0.02–6.20	0.6
4. Multistoried tree gardens	0.01–0.15	0.1
5. Tree plantations with cover crop/mulch	0.10–5.60	0.8
6. Shifting cultivation cropping	0.40–70.00	2.8
7. Agricultural intercropping in young forest plantations	0.60–17.40	5.2
8. Tree plantations, clean weeded	1.20–183.00	48.0
9. Forest plantations, litter removed or burned	5.90–105.00	53.0

Source: Wiersum (1984).

The impact of forest plantations also depends on the species planted. Villagers often find that *Eucalyptus camaldulensis* plantations tend to lower the water table and dry up shallow wells, as these forests can absorb up to ten times more water than pioneer forests and four times more water than secondary forests.

National and international attention has increasingly been paid to the possibility of combining tree species with field crops, i.e. agroforestry, for small landholders. The International Center for Research in Agroforestry (ICRAF), which is based in Kenya, has expanded its research activities to cover Asia. In Indonesia, about 70 percent of the total rubber is produced under agroforestry systems. These systems have been acknowledged as providing sustainable support in areas where the soils are too poor to grow food crops on a continuous basis. Agroforestry is estimated to have been adopted by about 7 million people and occupies approximately 2.5 million ha of land in Sumatra and Kalimantan, and includes the damar agroforest (Box I.3) (de Foresta and Michon, 1997).

Today, there are diverse and complex agroforestry systems that mix perennials with food crops. Some of the agroforests provide ecological functions similar to those of natural secondary forests, such as carbon sinks, sources of biodiversity, and means of alleviating soil erosion and flooding peaks (Garrity, 1998).

The development and adoption of agroforestry is largely dependent on various physical, environmental, political, social, and economic conditions. High population pressure combined with low per capita income and forest resources that are inadequate for local needs (including both timber and nontimber products) has necessitated the use of fuelwood and provides increased incentives for agroforestry. An abundance of State land under a strong land-use policy, together with government incentives and support, reduces the cost of and increases returns from agroforestry. However, the small amount of arable agricultural land per capita may limit the profitability of tree-based systems.

Agroforestry has been adopted and practiced widely in the PRC, India, Indonesia, Lao PDR, and Viet Nam. With the exception of Lao PDR, the population density in these countries is high, the income per capita low, and forest resources scarce. In India, for example, the annual requirement for fuelwood and timber is 220 million t and 280 million m^3, respectively, while the sustainable production levels of these products from the forests is only 30 million t and 12 million m^3, respectively (APAN, 1995). To meet the demand for such forestry products as well as to maintain sustained production from the forestry and agricultural sectors, agroforestry has been promoted in various parts of India.

The role of agroforestry in fuelwood supply is very important in both rural and urban areas of Bangladesh, where 90 percent of the fuelwood used is from home gardens. Agroforestry development in Lao PDR has been a response to government policy on land taxes and the desire of individuals to claim more land, especially land that is easily accessible.

In the PRC, agroforestry has the support of both central and local governments, with the aim of improving environmental and economic conditions. The systems used in the PRC are much more diverse than those in any other country in the region and include home gardens, strip shelter forests in combination with cropping systems, woodlot plantations, and tree shading with cropping systems. To address the shortage of fuelwood, more than 3 million ha of woodlot plantations have been established with an annual production of 10–30 t/ha, depending on the species planted. The shelter-forest system also increases wheat yields and latex production in rubber plantations. Further, it generates over 6 million m^3 of timber and 3 million m^3 of fuelwood per year through thinning and final harvesting.

The system in Lao PDR has been initiated mostly under the *taungya* system which involves planting teak or paper mulberry in combination with rice, pineapple, maize, or other cash crops to improve shifting cultivation. This is especially so in the northern part of the country where large shifting-cultivation areas are being replaced by teak.

Although agroforestry has been adopted and practiced successfully in many countries, the scale of its practice is limited to subsistence production. There are many factors that limit its growth and development, including insufficient mechanisms for the exchange of agroforestry experiences, inadequate policies, rules, and regulations, and an insufficiency or a lack of incentives from governments (APAN, 1996). Other factors include a lack of access to state-of-the-art technical information and farmer-generated knowledge among small households, a lack of up-to-date market information on agroforestry products, and inadequate support services for expanding agroforestry activities.

Livestock Subsector

Historically, livestock were raised using resources that were of little value for other uses, such as household food wastes or land that was not fertile enough for crops. Today, livestock production has emerged as one of the more advanced segments of agriculture, and also as one of the most important components of global agriculture. Of the three major production systems (land-based grazing, mixed farming, and industrial farming), industrial farming has seen the most rapid changes in the last decade, with most of the growth occurring in developing countries.

In Asia, where industrial and urban growth has been particularly rapid in the last decade, there has been a remarkable expansion of the peri-urban poultry and pig industries. This growth has been a response to increased demand as well as to technical advances. More concentrated feed and improved animal health have resulted in a more favorable feed-conversion ratio, giving these commodities a higher return on investment. The structural shift also reflects the pressure on land from population increase, rendering the expansion of land-intensive animal husbandry more difficult.

Livestock production is vital to the overall development goals of Asian countries: livestock production improves food security and nutrition and increases employment. Modern industrial farming operates side by side with small-scale farming, although there is still a predominance of mixed farming systems. Some 90 percent of livestock production in the developing countries of Asia comes from smallholders or landless persons.

Traditional farming is based on systems with minimal or no imported inputs and where livestock and crop activities are integrated. Farm products are mainly for domestic consumption and the excess is sold locally. The demand for livestock products has increased rapidly in urban centers, and traditional land-based livestock production with limited use of resources is unable to meet this increasing demand. The commercial nature of livestock production encourages specialized intensive farms to move nearer to the market place in the city, i.e. in peri-urban areas. Smallholder farms, because of their noncommercial nature, have little access to credit facilities or modern technologies to enhance their activities, further losing their competitiveness, resulting in their being supplanted and marginalized (FAO, 1998a). In some countries, rural producers, because of a lack of infrastructure, economies of scale, and insufficient marketing facilities, face heavy competition from urban producers, which often limits rural livestock production to subsistence levels.

In rural areas, there are many animals with a low level of productivity because they are only being fed at about maintenance level. With more feed, much of the additional nutrition would go directly to production. This has been demonstrated in Bangladesh where the provision of a small amount of supplemental nutrition has led to an increase in milk production from 1 to 6 liters per day in indigenous cows (Ramsay and Andrews, 1998). Increasing livestock production for poor farmers would provide a useful short- to medium-term benefit, especially where farm labor is underutilized. If animals already owned by the farmer become more productive, the benefit is received at little additional cost. Other constraints to livestock development in Asia for smallholders include the scarcity of feeds, high incidence of animal and poultry diseases, the prevalence of traditional livestock management systems, and inadequate access to credit.

Low livestock productivity can also be the result of poor animal husbandry. Appropriate livestock practices can make a contribution to raising productivity, but farmers often have limited access to education and training. Low productivity can also sometimes be attributed to the fact that these small-scale farmers do not have access to the better breeds of livestock. However, replacing local cattle with improved breeds will not solve the productivity problem unless feed resources on the farms are increased.

In the past, it has been government policy in many countries in Asia to focus on importing foreign breeds in order to achieve higher productivity levels. However, establishing a modern industry with imported breeds has proven difficult. This has been largely due to the fact that the greatest improvements that can be derived in animal productivity require better management and improvement of feed resources, not simply a different breed. Foreign breeds will most likely have different feed requirements that cannot be met by small-scale farmers and may not adapt to local conditions as well as do the indigenous breeds. Additionally, if feed requirements are not met for the foreign breeds, their reproduction rate may be affected.

Asia has relatively few major infectious disease problems in cattle and buffalo. However, foot-and-mouth disease has been difficult to prevent amongst countries with easily accessible land borders, unless joint programs have been implemented. In some Asian countries, notably Bangladesh, Indonesia, Philippines, and Thailand, the poultry sector is very important and the control of disease, especially Newcastle disease, is crucial, particularly at the village level. In Bangladesh, the estimated annual loss from poultry diseases is $240 million, of which about half is attributable to Newcastle disease alone. The increase in industrial livestock production has also increased the rate of livestock disease; diseases also spread more quickly and are harder to contain in intensive animal production.

In most developing countries, priority in the past has been given to the production of food crops. Several considerations highlight the current need to give greater priority to livestock development. While in the industrial world the demand for milk and meat will likely plateau, or even decline, in the developing world population growth and urbanization will fuel a strong increase in demand. For example, between 1986 and 1996, growth rates for meat in Asia were 7.7 percent, compared with only 0.7 percent for the rest of the world. Milk growth rates over the same period in Asia were 4 percent, whereas annual growth rates were negative at -0.8 percent for the rest of the world (FAO, 1998a). Current levels of milk and meat consumption in developing countries are only about one fifth of those in developed countries.

For the region as a whole, a strong trend has been observed towards the incorporation of more and more animal protein into the population's traditional vegetable-based diet. Connected with this trend is an increasing selectivity as to which parts of an animal are used for food. Traditionally, most parts of an animal were utilized, even if there was much wastage due to insufficient recovery and re-utilization technologies. Now the global trend is clearly to meat, and more often to lean meat. Other products, such as offal, blood, and bones, are increasingly used industrially and often recycled as feed.

The demand for livestock products is also highly income-elastic, and thus demand will increase with rising incomes. In Asia, the demand for meat is expected to triple by 2020 (de Haan, Steinfeld, and Blackburn, 1997). An accelerated livestock production program to satisfy this growing demand for primary livestock products such as milk, meat, and eggs will be required. Growth of all livestock sectors has been high over the past two decades (Table I.5), and this has been especially true of poultry and eggs. It will be a challenge for the livestock sector to satisfy future demands while at the same time preserving the natural resource base. The long-term objective is to produce and supply sufficient and safe animal protein for rapidly growing and urbanizing populations, under socially and environmentally acceptable terms.

Table I.5: Livestock Production in Asia

Livestock product	Production (t)				Annual Growth Rate (percent)		
	1965	1975	1985	1995	1966-1975	1976-1985	1986-1995
Total Meat	14,299,373	20,930,885	38,181,690	74,810,570	3.81	6.01	6.73
Pork	7,284,905	10,601,171	22,109,992	39,748,000	3.75	7.35	5.87
Poultry	1,512,179	3,078,570	6,027,808	13,951,590	7.11	6.72	8.39
Milk	44,975,411	58,064,570	89,094,320	142,617,600	2.55	4.28	4.7
Eggs	4,322,566	6,446,128	11,906,807	26,525,940	4.00	6.14	8.01

Source: FAOSTAT Database. *Available: http://apps.fao.org*

Growth in meat production until 2010 is expected to come from increases in productivity and from greater numbers of animals, with these factors accounting for about one third and two thirds of growth, respectively. Poultry production is projected to rise most rapidly, followed by pork production. More than 90 percent of the increase in pork production, however, will come from one region, East Asia, including the PRC. There are some differences between the major regions in structural changes, although the main trends are common to all. The proportion of poultry meat to total meat output is expected to continue to increase in every region while the contribution of cattle and buffalo meat will likely decline. Yields per animal are also expected to grow faster than in the past twenty years as a consequence of improvements in health, feed, and pasture carrying capacity.

Although the contribution of the livestock sector to GDP is relatively small, it is nevertheless significant in terms of the total output of the agricultural sector. The sector in Asia contributes from about 10 percent to about one third of agriculture's gross added value. However, livestock statistics generally quantify the products that are eaten and traded such as meat, milk, and eggs, and do not consider products such as draft power and manure (as fertilizer, fuel, or feed). As a result, the statistics greatly underestimate the role and importance of livestock (Ramsay and Andrews, 1999).

Growth rates have increased for all major livestock products over the past two decades. Productivity increases have been the main source of production increases, as opposed to the expansion of the industry or an increase in livestock numbers. Growth trends also differ within Asia, as there are subregional variations in production and consumption across the region. For example, South Asia consumes large quantities of milk but little meat. Farming is also still largely small scale with manure and draft power remaining highly important. East and Southeast Asia have seen the greatest increase in intensive production of monogastric animals such as pigs and poultry. In Thailand, poultry has become a major export earner.

The recent financial and economic crisis in Asia has also affected the industry, particularly in countries that rely on imported feed. These countries have seen a sharp increase in the prices of imported feed and other inputs as well as a contraction of demand. This has caused a drop in industrial production and in some cases it has been wiped out altogether. Industrial livestock production had grown rapidly in Indonesia until the crisis: large-scale poultry production, which was growing very fast there, was heavily damaged by the crisis, mainly because it is heavily dependent on imported feed. In Malaysia, although livestock production growth rates remain high, the industry also depends largely on imported feed.

Due to a lack of livestock feed resources and shortage of land for livestock feed production, most countries in the region are not in a position to develop large-scale intensive livestock industries without relying on imported feed (Ramsay and Andrews, 1998). However, some countries have turned to domestic feed resources. Thailand, for example, has now increased its use of cassava for livestock feed. Such developments offer a window of opportunity for the development of commercial smallholder livestock because the competitive pressure from large-scale industrial producers has, to a large extent, subsided.

In terms of total meat, less than 3 percent is produced under the grazing system. The bulk of meat production originates from mixed systems that account for two thirds of total production. Among the mixed farming systems, rainfed systems only account for one seventh, while the emerging industrial system now accounts for one quarter of total meat production (FAO, 1998a). About 21 percent of the world's arable land is producing feed for the livestock industry, and livestock production uses 32 percent of total cereal production. Maize accounts for half of all feed grain, with barley and wheat as the other main components. Soybean is the most important component among oilmeals.

The most remarkable growth trends have been in the PRC, Indonesia, Malaysia, and Thailand (Table I.6). The growth rate in total per capita meat production in Asia almost doubled from an average of 3.81 percent annually during 1966–1975 to 6.73 percent during 1986–1995. The PRC experienced the strongest growth rates for meat production, followed by Malaysia. Annual growth in pork production slowed slightly from 1976–1985 to 1986–1995 from an average of 7.4 percent to 5.9 percent (Annex Table A27). The PRC has the highest pork production per capita in Asia, followed by the Republic of Korea, the Philippines, and Viet Nam, and strong growth was seen only in these countries. Cambodia has maintained relatively fair growth rates, and is now experiencing moderate levels of pork production per capita.

Milk production shows a similar trend to that of meat production, although at lower rates (Annex Table A28). The highest per capita production is in South Asia, where India and Pakistan are leading, although growth in the industry is now slowing there. The PRC and Thailand have also shown strong growth rates, but per capita production is still low. Strong growth has continued for poultry and egg production, averaging 7 percent or more per annum during the last 30 years (Annex Tables A29 and A30). Malaysia and Thailand have had strong growth in both areas, and although the PRC has also shown strong growth rates in both areas, production per capita of poultry is still low. Japan has high egg production per capita and moderately high poultry production per capita, but growth rates are low because local demand is satisfied by imports, and the industry is now stabilizing.

Areas of South Asia have experienced strong growth in livestock production over the past 30 years. Strong growth in poultry has been manifested in India, Bangladesh, and Pakistan (Annex Table A29). In India, livestock production is a high priority at every administrative level for two primary reasons: production self-sufficiency and rural development. Small farm size remains the biggest obstacle to development of the livestock industry there.

Meat production has also shown a strong growth trend in Pakistan (Table I.6). Pakistan is particularly self-sufficient in livestock products. Although this country is the largest milk producer in the region, some milk powder is still imported due

Table I.6: Total Meat Production Per Capita in Asia

	Production (t)			Production Per Capita (t per 1,000)			Annual Growth Rate (percent)		
	1975	1985	1995	1975	1985	1995	1966–1975	1976–1985	1986–1995
ASIA	20,930,885	38,181,690	74,810,570				3.81	6.01	6.73
East Asia									
China, People's Rep. of	9,762,314	20,624,240	47,752,610	10.5	19.3	39.1	3.81	7.48	8.40
Japan	2,213,762	3,460,969	3,200,840	19.9	28.6	25.6	7.29	4.47	-0.78
Korea, Rep. of	249,120	735,408	1,416,683	7.1	18.0	31.5	5.71	10.82	6.56
Mongolia	229,060	226,000	214,427	158.3	118.4	87.1	4.16	-0.13	-0.53
Southeast Asia									
Cambodia	52,320	87,728	153,508	7.4	11.8	15.3	1.68	5.17	5.60
Indonesia	530,621	1,002,546	1,936,497	3.9	6.0	9.8	3.35	6.36	6.58
Lao PDR	18,370	35,554	49,141	6.1	9.9	10.1	-3.42	6.60	3.24
Malaysia	243,003	420,607	956,259	19.8	26.8	47.5	6.74	5.49	8.21
Myanmar	199,279	311,993	335,467	6.5	8.3	7.4	3.92	4.48	0.73
Philippines	603,416	704,408	1,622,850	14.0	12.9	23.9	2.68	1.55	8.35
Singapore	113,965	131,135	147,872	50.4	48.4	44.4	8.76	1.40	1.20
Thailand	657,438	1,074,149	1,473,500	15.9	21.0	25.3	4.66	4.91	3.16
Viet Nam	424,140	864,563	1,385,620	8.8	14.4	18.8	-0.20	7.12	4.72
South Asia									
Afghanistan	203,428	217,760	230,520	13.2	15.0	11.7	2.86	0.68	0.57
Bangladesh	239,957	259,907	370,837	3.1	2.6	3.1	1.72	0.80	3.55
Bhutan	4,613	6,544	7,764	4.0	4.5	4.4	2.48	3.50	1.71
India	2,238,951	3,077,774	4,391,485	3.6	4.0	4.7	2.15	3.18	3.55
Maldives	565	750	850	4.1	4.1	3.3	2.50	2.83	1.25
Nepal	101,015	166,456	204,648	7.9	10.1	9.5	3.60	4.99	2.07
Pakistan	564,780	946,774	1,856,250	7.6	9.4	13.6	3.32	5.17	6.73
Sri Lanka	58,176	55,373	88,108	4.3	3.4	4.9	1.12	-0.49	4.64
Central Asia									
Kazakhstan			1,065,236			63.3			
Kyrgyz Republic			179,900			40.3			
Uzbekistan			519,000			22.8			
Tajikistan			56,400			9.7			
Turkmenistan			110,500			27.1			

0 = zero or less than half of the unit measured.

Note: Annual Growth Rate = ((Ln(value year begin) - Ln(value year end)) / number of years) x 100.

Source: FAOSTAT Database. *Available: http://apps.fao.org*

to the lack of a proper fresh-milk collection network in the country. Live animals, hides, and skins are exported. The country has buffalo and cattle breeds for the production of milk whose quality is world famous. There is still much room for improvement, and it is estimated that the output of livestock products could double if extension services were improved. Among the constraints are an inadequate feed base and a low number of productive animals.

In Bangladesh there is a large and important livestock population, but the animals have low productivity, mainly due to the insufficient quality of feed resources and a high incidence of disease. Bangladesh is deficient in livestock production but cannot afford many imports. However, growth of the industry has been improving over the past 30 years.

Production has remained low in other countries in South Asia such as Nepal, where 90 percent of the population depends on integrated farming, including livestock raising. Feed resources need to be developed, which would lead to more productive animals. Difficult transportation conditions and weak marketing facilities are among other constraints facing Nepal today. However, egg and poultry production have shown moderate increases in growth rates since 1975.

Central Asia has the highest per capita production of milk and a high production of meat per capita. However, the production of other livestock products is low. Data have not been available to determine growth rate trends over the past three decades for Central Asia.

Livestock production has a great potential for contributing to either the degradation or the enhancement of the environment, depending on the technologies and practices adopted. The environmental challenges range from overgrazing (degradation and erosion of land), to deforestation, to degradation and pollution of water resources, emission of greenhouse gases, and loss of biodiversity. Growth rates in the more developed countries such as Hong Kong, China; Japan; and Singapore have begun to level off. Land scarcity in Hong Kong, China; and Singapore dictates that they will remain largely importers of livestock products. Recently, the governments in these countries have paid more attention to environmental pollution and health risks resulting from livestock farming than to increasing production capabilities.

The threat to human health from livestock production is exemplified by the spreading of a rare virus from pigs to humans in Malaysia in early 1999. More than 100 people died and almost one million pigs were destroyed to keep the virus from spreading.

Recent rapid growth of the livestock sector near urban areas has created additional pollution problems for already congested and relatively highly polluted areas. Wastes generated by intensive production units, especially pig production, are a major source of water, land, and air (odor) pollution. Heavy metals arising from feed can affect the health of nearby residents if wastes are not treated properly. The use of additional water to clean solid waste from production units further increases the amount of waste to be treated and in turn the cost of wastewater treatment.

Animal waste from intensive systems can be used to produce biogas for heating, drying, and power generation. The digester effluent can also be used to fertilize crops and fertilize/ feed fishponds. Appropriate government intervention in this area would allow environmental problems to be solved and maintain long-run production, while deriving benefits from biogas energy and nutrient recycling (FAO, 1998a). Other initiatives, such as limitations on stocking density, would help to reduce pollution discharges.

Pressure for greater emphasis on environmental conservation is evident throughout the region. The greatest threat is from overgrazing and in many countries, especially in semi-arid areas, livestock numbers already exceed the carrying capacity of unimproved natural grasslands. Overgrazing is a major concern, especially in India and Central and inner Asia. It will be important to identify situations where the raising of livestock is out of balance with the adsorptive capacity of the soil, water, and air. Competition between crops and grazing generally results in resource degradation and finally in the collapse of livestock production, especially for the larger ruminants. In some areas, there is a need to restrict the density of animals raised, the objective being to optimize the long-term productivity of the land as a whole while maintaining ecological diversity and environmental balance.

In tropical Asia, the problem of deforestation is complicated as it is intertwined not only with livestock production, but also with logging operations and human population pressure, especially in areas where suitable land is scarce. Overgrazing, which leads to land degradation, also leads to deforestation as new land is cleared for use.

Of the livestock production systems, mixed farming tends be the most environmentally friendly: it allows reuse of animal waste as organic fertilizers that replenish the soil and reduce erosion. Mixed farming also provides protection against product/price fluctuation risks. Each year, livestock produce about 13 billion t of waste, and supply 22 percent of total nitrogen fertilizer and 38 percent of phosphates of animal origin (FAO, 1997a). Crop waste can also be reused as animal feed.

In livestock systems other than mixed farming, transport costs from nutrient-surplus to nutrient-deficient areas can be high. For example, with intensive systems, there is a need for transport of feed to and manure from production areas. It may eventually be less costly to relocate intensive production areas away from peri-urban centers, if a proper transportation infrastructure can be developed. Some rapidly expanding urban centers such as Ho Chi Minh City, Viet Nam, have already initiated programs to move intensive pig and poultry production outside city boundaries. The difficulties in transition from extensive to more intensive livestock production have also led to increased environmental degradation.

There have been massive investments in past decades in physical infrastructure, including the construction of roads, ports, and communication facilities, but also of slaughterhouses, and cold storage and retail facilities. All of these have greatly reduced transaction costs and have, in numerous places, made possible trade in livestock products. Because of the perishable nature of livestock products, infrastructure development has an extremely stimulating effect on livestock production.

As per capita incomes increase, more animal products pass through market and processing channels before consumption. This leads to even greater waste production. The most important environmental impact of animal product processing results from the discharge of wastewater. Additionally, heavy metals such as copper, zinc, and cadmium are used as growth stimulants in some feeds. Without proper management of these discharges, intensive farming systems discharge waste containing heavy metals at levels that are harmful to animals and human health. At present, only the Organisation for Economic Co-operation and Development has regulations that aim to reduce the level of heavy metals in feeds.

Fisheries

Asia accounted for more than half of total world fishery production in 1996 (ICLARM, 1999). During the two decades ending in 1996, total production of fish and shellfish in Asia increased at a much faster pace (approximately 4 to 5 percent per year) than did the production of food crops, rising from nearly 28 million t to 67 million t from 1975 to 1996, raising Asia's share of the world total from 42 percent to 55 percent (Table I.7). The overall performance of Asian fisheries is especially remarkable when their growth is compared with that of the rest of the world. Asian fishery production grew at an average 4.8 percent per year over the past decade, up from 3.6 percent a decade earlier, whereas that of the rest of the world declined from 2.6 to 0.3 percent over the same period (Table I.7). Asia was the main force propelling the overall increasing trend in world fishery production.

Two striking features of Asia's fisheries growth during the past decade are the emergence of the PRC as the predominant producer, and an increasing contribution from aquaculture. The PRC contributed nearly half of Asia's total fish and shellfish production in 1996, compared with only 16 and 21 percent in 1976 and 1986, respectively. Most of the remaining production was contributed by Japan, India,

Table I.7: Asia and the World: Fish and Shellfish Production,
Selected Years

	Output (t, million)						Growth (%)	
	1950	1975	1985	1990	1995	1996	1977–1986	1987–1996
Asia	6.46	28.15	38.47	47.45	63.28	67.11	3.6	4.8
Rest of the World	12.74	37.59	48.67	51.56	54.00	53.90	2.6	0.3
World Total	19.20	65.74	87.14	99.01	117.28	121.01	3.0	2.6
Asia's Share in world (%)	33.7	42.8	44.1	47.9	54.0	55.5		

Source: From data presented in FAO (1998b).

Indonesia, Thailand, Republic of Korea, and the Philippines
(in that order), each of which produced over 2 million t on
average during the past decade (Annex Table A31). However,
only a few have registered annual growth rates above 3 percent,
while production in Japan and the Republic of Korea has
declined. Japan, Asia's largest fish producer until 1988, now
takes second place after the PRC. Japan's share in Asia's total
fish and shellfish production dropped from 34 percent in 1976
to merely 10 percent in 1996.

Much of the growth of Asian fisheries during the past
decade was fuelled by aquaculture production, which grew by
over 11 percent per year between 1987 and 1996 in both
freshwater and marine waters (Table I.8). Once again, most of
this growth has taken place in the PRC, where fish production
increased by a factor of 6.6 during 1977 to 1996, with particularly
high growth, an annual average of 13 percent, during the past
decade (Annex Table A31). Over half of the PRC's fish and
shellfish production in 1996 came from aquaculture, and
accounted for 75 percent of Asia's aquaculture production of
fish and shellfish, up from 55 percent in 1986. The PRC is
probably the only country in the world where culture production
exceeds capture production (Lan and Peng, 1997). Aquaculture
has also spread rapidly in other Asian economies, notably India,
Indonesia, Japan, Republic of Korea, Philippines, and Thailand.
However, in many of these countries, the commercialization
and consequent intensification of aquaculture, the use of

carnivorous species that depend on fishmeal extracted from capture fisheries, and the negative environmental and socioeconomic impact, have raised many doubts about its overall benefits and sustainability.

Asia has a long history of capture and culture fishery production, evident in the wide variety of traditional gear and culturing techniques that have evolved over time to exploit the diversity of resources. Fish form an important part of the diet of many Asians, although per capita consumption varies widely from country to country. It is generally low in land-locked countries, such as Bhutan, Kyrgyz Republic, Mongolia, Nepal,

Table 1.8: Fishery Production in Asia, 1976–1996

	Production (t)			Average Annual Growth Rate (%)	
	1976	1986	1996	1977–1986	1987–1996
Total (fish, shellfish, aquatic plants)	30,787,197	45,218,268	75,158,690	3.8	5.1
Fish, Shellfish	28,963,046	41,691,808	67,112,800	3.6	4.8
Aquatic Plants	1,824,151	3,526,460	8,045,890	6.6	8.3
Inland Fisheries (fish, shellfish)	4,122,941	7,869,782	19,403,434	6.5	9.0
Marine Fisheries (fish, shellfish)	24,840,105	33,822,026	47,709,366	3.1	3.4
Capture Fisheries (fish, shellfish)		34,447,940	43,647,733		2.4
Aquaculture (fish, shellfish)		7,243,868	23,465,067		11.8
Marine Capture (fish, shellfish)		31,424,480	38,899,011		2.1
Inland Capture (fish, shellfish)		3,023,460	4,748,722		4.5
Freshwater Culture (fish, shellfish)		4,526,454	14,177,520		11.4
Brackishwater Culture (fish, shellfish)*		590,105	1,218,798		7.3
Marine Culture (fish, shellfish)		2,398,956	8,810,994		13.0

Sources: From data presented in FAO (1998b, 1998c).

Note: * Brackishwater culture includes both inland and coastal waters. In 1996, the two represented, respectively, 39% and 61% of total brackishwater production, with average annual growth rates (1987–1996) of 4.0% and 10.1%, respectively.

and Tajikistan, and in South Asia, but relatively high in Southeast and East Asia, particularly among wealthier economies.

The contribution of this sector to food security is highlighted by the fact that increased supplies of fishery products in the PRC raised the annual consumption of aquatic products there from 2.67 kg to 7.29 kg per capita during 1952 to 1992. This is especially significant given that the PRC population more than doubled in that period from 575 million to 1,172 million (Wang, 1996, cited in Williams and Bimbao, 1998).

The fishery sector in Asia employs a large workforce. Even though the proportion of people dependent on fisheries in Asia might appear small against the region's vast population, their number is considerable in absolute terms. According to FAO (1998d) some 25 million fishers and fish farmers–four fifths of the world total–are employed in Asia. In South and Southeast Asia, fisheries employ 10.36 million people as full- or part-time fishers, with 8.64 million employed in marine fisheries and the remaining 1.72 million in inland fisheries (Hotta, 1996). Moreover, there may be a large number of occasional fishers, particularly in the inland fisheries. In addition to the direct employment provided by fisheries, considerable job opportunities exist in the related service and transport industries. Opportunities for women exist especially in aquaculture, fish retailing, and processing. In the PRC alone, the population engaged in these fishery-related activities (not including capture fisheries and aquaculture directly) numbers over 11 million (Bureau of Fisheries, Ministry of Agriculture, PRC, 1997).

Marine capture fisheries, which contribute the largest, albeit declining, share of Asia's total fisheries production, are characterized by the presence of a large number of small- and medium-scale fishers (Hotta, 1996) (Box I.4). They operate in shallow inshore waters of up to 50 m in depth, using traditional but increasingly modernized craft and gear as well as nontraditional craft and gear. Fishing pressure is intense along the continental shelves of the western coast of India, the Bay of Bengal, the Gulf of Thailand, the South and East China seas,

Box I.4 Multiple-Use Conflicts and Asian Fishers

Despite the technical advances made in other sectors in many Asian countries, the majority of Asia's fishers are small-scale coastal fishers who are generally among the poorest of the poor, and for whom the open-access nature of fisheries offers a last resort to eke out a living. For many of these people, the largest proportion of their income is spent on food and a large proportion of their food comes from coastal fisheries.

A diverse group of other stakeholders coexists with coastal fishers. These include the trawler operators, whose fishing gear is highly destructive, disturbs habitats, and harvests indiscriminately for the fishmeal industry. Commercial fishers with better equipment often compete for resources in the same fishing grounds as coastal fishers. The destruction of mangroves and water pollution from aquaculture both serve to reduce the regenerative capacity of coastal areas. Depleted fish stocks are the result of competition, coastal trawling, and environmental degradation, and pose direct threats to the livelihood of all fishers. It is not surprising that conflicts between small- and large-scale fishers are widespread (Silvestre and Pauly, 1997) and likely to increase.

A number of policy options are available to improve the livelihood and/or productivity of small-scale fishers. On the technical side, stock enhancement technologies could be introduced. Institutional solutions, such as rights for marine aquaculture and participatory management by local communities, may be needed to replace the lax enforcement of laws on trawling, for example. Protection afforded to the fishmeal industry could be scrapped. Alternative employment opportunities also need to be created. The large number of coastal fishers suggests that, for some, a transition away from small-scale fishing may be necessary.

In boom periods, a transition away from low-productivity small-scale fisheries is easier because there is increased demand for labor from other sectors. In Thailand, for example, the economic boom of the late 1980s to mid-1990s created new

(continued next page)

Box 1.4 (continued)

employment opportunities either outside the fishing sector or in other fishery subsectors such as aquaculture. Although the total number of people employed in marine fisheries seems to have declined relative to total population, there has been a large internal change, with traditional fishers moving out of the industry. In place of these traditional fishers are newcomers from the peripheral regions of the country (the northeast and upper north), or even from neighboring countries (Myanmar, Lao PDR, and Cambodia) where few income opportunities exist (Boonlert, 1994, cited in Mingsarn and Pednekar, 1998).

Such a high turnover is possible when economic growth creates alternative employment opportunities. When no such alternatives exist, however, fishers are left with no choice but to compete for a dwindling resource, leading to further overfishing, resource degradation, and finally untenable social conflicts. Unless drastic action is taken to strengthen and enforce responsible coastal fishing, coastal fisheries, which have been valuable assets for many Asian countries, will diminish and turn into social liabilities in the near future.

the Bohai and Yellow seas, and the parts of the Sea of Japan bordering Japan.

Only a handful of nations, notably Japan, Republic of Korea, Democratic People's Republic of Korea, and more recently the PRC, have sizeable fleets of large vessels (over 100 gross registered tons) (FAO, 1997c), allowing them to fish further offshore and in high seas. Following the 1982 Convention on the Law of the Sea, and the declaration of 200-mile exclusive economic zones (EEZs), many countries have enhanced their ability to fish further offshore. A number of South and Southeast Asian countries, such as India, Indonesia, Philippines, and Thailand, have begun developing their distant-water fishing fleets. These mostly operate in the EEZs of other countries under bilateral agreements.

The gross statistics of Asian fisheries might appear to contradict the general picture of ill health in most major fisheries of the world. Many of the world's major fisheries are facing serious falls in production; stocks of a number of commercially important fish species have been fully or overexploited or are rapidly dwindling, and many commercial species are endangered. The trend to fish "down the food chain" may have long-term and perhaps irreversible impact on the ecological balance in marine ecosystems (Pauly et al., 1998). The open-access nature of fisheries and "subsidy-driven over-capitalization" have been largely responsible for the trends of overfishing and excess capacity that have led to a global crisis in fisheries (Garcia and Newton, 1997, cited in Pauly et al., 1998). These trends already exist in many Asian waters, casting shadows on the sustainability of fisheries growth in Asia.

The predominance of a single country such as the PRC makes sustainability of Asian fisheries growth even more vulnerable because it hinges largely on the ability of the PRC to sustain its high production rates, particularly in aquaculture. However, even though further potential for aquaculture expansion in the PRC has been identified (e.g. ADB, 1995b), the two major constraints for aquaculture generally, viz. environmental degradation (from aquaculture itself as well as from external sources), and competition for land and water resources from other economic activities (FAO, 1997b), are also likely to threaten the goal of realizing that potential. Intensification of aquaculture has been contributing to further environmental degradation, which is already apparent in the PRC and elsewhere in Asia (FAO, ibid.).

There are several indicators that suggest that overfishing in many parts of Asia may have worsened during the last two decades. Demersal stocks have been heavily fished in much of the Gulf of Thailand since the 1970s. In Thailand, catch rates in terms of catch per unit effort are currently only 6 to 10 percent of their peak levels, which were reached soon after the introduction of otterboard trawling in the early 1960s (FAO, 1997b; Mingsarn and Pednekar, 1998). Catches of a number of large and small pelagic stocks also appear to have declined, although the recent decline in landings of small pelagic fish is largely attributed to the environment-linked fluctuations in catches of the Japanese pilchard in the northwestern Pacific (FAO, 1997c). More importantly, however, catches of miscellaneous species, which traditionally form a large part (nearly a quarter) of Asia's marine capture fisheries, have been on the rise (Annex Table A32). A considerable and increasing portion of these catches consists of juveniles of commercially important fish species. An environmental and natural resource accounting exercise carried out in the Philippines estimated the natural resource depreciation of fisheries in 1992 at 6.5 billion pesos (at 1988 prices), higher than that of forests and soils by a factor of 13 and 11, respectively (IRG/Edgevale/REECS, 1996).

A recent study analyzing the world's top 200 marine fish resources, indicated that in 1994 about 35 percent registered declining landings and thus were in the senescent phase. Another 25 percent were in the mature phase at a high level of exploitation, while the remaining 40 percent were still developing (Grainger and Garcia, 1996). Thus, 60 percent of the world's fisheries (including some in Asia) are in urgent need of management to control and reduce fishing capacity and effort (Grainger and Garcia, ibid.). For instance, large increases in the capture of marine cephalopods and other molluscs are attributed to a decline in demersal fish, which are their predators. Catches of these molluscs may increase in the short term, since the prey population is much larger than that of the predators, but the value per unit catch decreases and the increasing catches give "a misleading vision of the state of world fishery resources and a false sense of security" (Grainger and Garcia, 1996, cited in FAO, 1997b).

Capture Fisheries

The declining share of marine capture fisheries to total fishery production in Asia is mostly attributable to slower growth in the northwestern Pacific, traditionally dominated by the commercial fisheries of PRC, Japan, Republic of Korea, Democratic People's Republic of Korea, and nonAsian countries such as Russia, the USA, and several European nations. Yet, the average production growth rate of 1.9 percent per year in this fishing area during 1987 to 1996 was still higher than the world average of 1 percent (Annex Table A33). The drastic decline in recent years in the catches of the two dominant species, viz., the Japanese pilchard (sardine) and Alaskan pollock, has been the main reason for the slower growth of the northwestern Pacific capture fisheries. Large increases in the catches of other species, especially Japanese anchovy, largehead hairtail, Japanese flying squid, and salmon could not fully compensate for the losses from these two species.[3] Coastal fisheries in this region, particularly in the seas bordering the PRC (Yellow Sea, Bohai Sea, East China Sea, and South China Sea) are constrained by poor water circulation caused by the semi-enclosed nature of these seas (Deb, 1997).

Marine capture fisheries still account for the largest share of Asian as well as global fish production. The share of inland capture fisheries has been marginal. These fisheries have grown at twice the rate of marine capture production during the past decade, in part due to the large number of reservoirs constructed and increasing efforts to seed inland waterbodies. However, further expansion may be constrained by environmental degradation and the fact that most inland waterbodies have been already exploited.

The increased mechanization of traditional craft and the development of new gears have spurred more intensive fishing further offshore, leading to reduced catches despite increased fishing effort. For example, catches of pelagic fishes in much of the Gulf of Thailand declined within several years of the introduction of purse seines in Thailand in the early 1970s (Mingsarn and Pednekar, 1998). Some fishing technologies, e.g. mechanized trawlers and purse seiners, were introduced in India

---

[3] While the decline in Alaska pollock catches may have been due to overfishing, that of Japanese sardine seems to be caused by decade-long changes in the marine environment, and to correlate with the El Niño-Southern Oscillation (ENSO) index (FAO, 1997b).

somewhat later than they were in Southeast Asia (Devaraj and Vivekanandan, 1997). Unlike their Southeast Asian counterparts, local Indian craft did not really begin to be motorized until the 1980s. Nevertheless, there has been a rapid increase in their number in India in recent years: 47,000 such vessels in total according to recent estimates (Gopakumar, 1997). Mechanization has increased the ability of small-scale Indian fishers to fish further, up to 100 nautical miles, from shore. One consequence of this has been increased opposition from these small-scale fishers to the operation of foreign fishing fleets in India's EEZ (Gopakumar, ibid.), which begins 12 nautical miles from the eastern coast and 24 nautical miles from the western coast.

Similarly, since the introduction of otterboard trawling in Asia, demersal fishes have been overfished or heavily fished in most shallow coastal waters. Perhaps the only major exception is in the western Indian Ocean, where demersal production has steadily increased since the 1950s, especially of croakers and drums since the early 1980s (Devaraj and Vivekanandan, 1997).

Tunas have assumed greater importance, particularly in the western Central Pacific, following the rapid development in the early 1980s of purse seine fisheries in Southeast Asia for canning, and since the mid-1980s, of longline fisheries targeting tuna for the fresh *sashimi* (raw fish) market (FAO, 1997b). Significant catches also come from the eastern Indian Ocean and the northwestern Pacific. Although catches appear to have peaked recently, large potential resources may exist in the western Indian Ocean, given the strong upwellings in northwestern areas (FAO, 1997b).

The greatest growth in the past decade, however, has been registered not by catches of marine finfish, but by catches of diadromous fish such as salmon, shads, and trout, and by a variety of marine invertebrates (crustaceans, molluscs, echinoderms, and miscellaneous marine invertebrates) (Annex Table A34). The significant increases in catches of cephalopods and other molluscs are mainly due to more intensive fishing of these lightly exploited resources; however, their abundance may also result from the depletion of demersal fish, their predators, due to overfishing. Demersal fish stocks have been overfished since the 1970s in most areas, particularly the East and South China seas, the Yellow Sea, and the Gulf of Thailand.

Overfishing not only poses further threats to dwindling fish stocks and conservation efforts, but probably has also caused changes in the marine ecosystem. However, the exact nature of the predator-prey relationship between different species and its impact on stock abundance is not very well understood. It is, therefore, difficult to estimate, for instance, how much of the current upward trends in the landings of pelagic fish (as well as cephalopods) are due to the depletion of predatory demersal fish. It is also impossible "to determine to what extent the rehabilitation of the overfished demersals will affect the survival and potential of the pelagics" (FAO, 1997c).

The fishery industry, dominated by capture production, is among the few remaining frontiers of hunting and gathering in human society. The pressure exerted on the world's fishery resources, owing to modern fishing technology and lax resource management, has been so intense that in most cases making fisheries sustainable would require control of access and reduction of fishing effort on overexploited resources, including demersals and straddling stocks (i.e. of species whose range includes the waters of two or more countries). These measures would have to continue until some recovery, indicated by increased catch rates, became evident. Management efforts should therefore focus on sustaining rather than increasing catches from marine fisheries (Devaraj and Vivekanandan, 1997).

Largely due to population pressure on inland water resources, inland capture fisheries in Asia have not developed on a large scale, contributing only about 7 percent of total fish and shellfish production. Although inland capture production in Asia grew at the moderate rate of 4 percent during the last decade, its share in total production declined slightly. Worldwide, inland capture production has stabilized at around 6.5 million t after peaking in 1990, and this level is expected to be maintained until at least 2010, although wide local variations may occur (Coates, 1995). The major constraints to inland capture fisheries are the growing threat from pollution due to increasing urbanization and industrialization, and inadequate access and user rights (Coates, 1995).

It should be noted, however, that a significant part of inland capture landings does not enter into official statistics, because some of the harvest is consumed directly by fishers and their families. In Thailand, for example, such direct consumption is estimated at 25 percent of the reported catch (FAO, 1997c).

Realizing the importance of inland fisheries (both capture and culture) to rural food security, intensive aquaculture and culture-enhancement techniques are being used in many Asian countries to increase productivity through seeding or stocking of waterbodies. With the further development of hatchery technology, these practices should offer greater potential for realizing higher yields. Overall, culture-enhanced capture fisheries seem to offer better potential for low-income, resource-poor communities, because they use existing water resources and low resource-input systems, and create little, if any, pollution. They may thus be better suited than intensive aquaculture for rural communities.

Aquaculture

Records of aquaculture practices in Asia date back to the manual of fish culture written by Fan Li in China some 2,400 years ago (Deb, 1997). Long histories of aquaculture practices are also evident in a number of other Asian countries, such as polyculture of freshwater fishes in natural and human-made ponds and tanks in India and the tambak culture system in Indonesia. These traditional systems are mainly extensive forms of culture with little, if any, external inputs such as feed.

The rapid spread of aquaculture and its diversification in many Asian countries is due largely to successes in hatchery technology and balanced-diet feed manufacturing. These and various other technologies have allowed intensification of aquaculture, increasing productivity. Today, Asia is the largest producer of cultured fish, contributing almost 90 percent of the world culture production of 25 million t in 1994 (ICLARM, 1998). Particularly high growth rates have been attained for freshwater finfish culture (carps, tilapia, and a number of other species) and the culture in marine waters of cephalopods (squids, octopuses, cuttlefish) and oysters, for which technological breakthroughs have been achieved and adopted all over Asia.

Marine aquaculture has also grown at a fast pace. Its share of aquaculture rose from 33 to 37.5 percent during 1987–1996, while that of freshwater culture fell from 62 to 60 percent. In terms of value, however, brackishwater culture had a larger share (21 percent on average during 1987-1996), than its small volume (7 percent) would otherwise indicate. Although brackishwater culture overall grew relatively slowly, coastal aquaculture, particularly of penaeid shrimp, but also of cephalopods, grew at a high rate, especially for several major exporters such as Bangladesh, PRC, India, Indonesia, Thailand, and Viet Nam. In contrast, inland brackishwater culture declined in some countries, such as the Philippines and Thailand, and grew rapidly in others, particularly Bangladesh and Taipei,China.

Inland production grew at the rapid rate of 11 percent per year during the last decade, and its share in total fish and shellfish production increased from 19 to 29 percent. Much of this growth was contributed by fresh- and brackishwater finfish aquaculture, particularly in Bangladesh, PRC, India, and Indonesia. The PRC has recently become the major player in inland fishery production. Inland fisheries production there grew by a factor of 11.6 during 1977–1996, and its share of world inland production increased from less than 16 percent to 40 percent in the same period[4].

Much of the aquaculture growth in Asia is due to the PRC's great increases in freshwater carp culture, marine aquaculture of mussels and other molluscs, and the culture of algae such as kelp and laver, as well as brackishwater shrimp culture (ADB,

---

[4] From data presented in FAO (1998b).

1995b). The PRC's successes in inland culture, particularly of carp, are being repeated elsewhere. The PRC and India, in particular, have shown remarkable growth in aquaculture over the past decade, averaging 15 and 9 percent annually, respectively. These two countries accounted for 83 percent of Asia's total aquaculture production. The proportion of PRC freshwater production in total production is significantly higher than that of other countries. In India, freshwater culture has been growing by 6 percent annually and produces 1.6 million t per year from the current 800,000 ha of culture area. According to the head of one of India's leading aquaculture research institutes, the current area can be extended by an additional 2.2 million ha to meet the estimated potential demand of 4.5 million t (FAO, 1998e).

Growth in aquaculture is likely to continue, since much potential for inland aquaculture remains untapped in many countries, including the PRC, and new technology, particularly in genetic improvement and hatchery operations and rearing, continues to be developed. However, the ability to realize this potential also rests on a number of factors, such as the rehabilitation of degraded or polluted water resources, the development of and access to markets, price incentives, and processing facilities, as well as minimizing the environmental impact arising from aquaculture operations. In the PRC, the further utilization of waterlogged areas and rice fields, and the rehabilitation of ponds offer high growth potential (FAO, 1997c).

The fast growth of aquaculture is a response to both high- and low-end markets. Coastal shrimp culture has developed in response to increasing demand from international high-income markets and the resultant price increases, whereas freshwater finfish aquaculture is usually focused on low-value food fish. For the former, the income earned has been illusory and deceptive because neither the farmers nor the governments have fully considered the overall cost of shrimp farming (Box I.5). The boom in freshwater low-value food-fish aquaculture, by contrast, has benefited the poor, because fish is a major source of protein for a large number of Asian countries, including Bangladesh, PRC, Indonesia, Myanmar, and Thailand.

Box I.5 Shrimp Aquaculture

Since the mid-1980s, the culture of penaeid shrimp has increased dramatically in many economies of Asia, particularly in South and Southeast Asia. The main species cultivated in Asia is *Penaeus monodon* or black tiger prawn, due to its rapid growth, relatively large size, and increasing demand from world markets. Cultured shrimp production now accounts for over 30 percent of global shrimp production. Asia produced nearly 80 percent of the world's cultured shrimp; Latin America most of the rest.

Shrimp aquaculture is characterized by cycles of "boom and bust". In Taipei,China, where the intensive technology was first developed in the early 1980s, a series of production crashes occurred after a period of spectacular yields during the late 1980s (Csavas, 1992). Similar patterns of boom and bust followed elsewhere, most notably in Bangladesh, PRC, India, Indonesia, Thailand, and Viet Nam. Disease and poor crop and environmental management, especially of water quality, were the main causes of these failures. Farmers have overcome these problems by "shifting cultivation", i.e. moving to relatively pristine areas.

Intensive shrimp culture, driven by lucrative demand from overseas markets, has created adverse socioeconomic and environmental impact, not only to third parties but also to the shrimp industry itself, and which has raised questions about its sustainability. Among the major environmental consequences of shrimp farming are the destruction of mangroves, water pollution from pond discharges, land dereliction, saltwater intrusion to adjacent nonshrimp farms, and the introduction of exotic species and diseases in coastal waters (Briggs, 1994, and Dierberg and Woraphan, 1996, cited in Direk et al., 1998). Shifting cultivation further aggravates and spreads these pathogens and their impact to new locations. The introduction of shrimp farming has apparently affected rice and orchard farmers, whose production is suffering from increased soil salinity. Many shrimp farmers have themselves fallen into debt due to an outbreak of disease following a short-lived spell of prosperity.

Because the above social and environmental costs have not been internalized, the exporting countries have actually

(continued next page)

Box I.5 (continued)

underpriced their products. Many countries have started to adopt policy measures to limit shrimp farming and to reduce its environmental impact. The efforts to date have been either too little or too late. The prospect of making quick profits continues to attract more people to this activity, and cost considerations and concerns for profitability make them disregard most environmental regulations. Valuable lessons on curbing environmental impact could be learned from the experience of salmon farming in industrialized nations, where regulatory frameworks and strict enforcement have helped to limit the social costs of aquaculture and made the subsector truly high yielding.

The growth of freshwater aquaculture in these countries has substantially raised the nutritional level of the poor.

Tilapia, a group of fish of African origin, is emerging as a rising star of freshwater aquaculture in Asia. Between 1988 and 1996, the production of tilapia increased fourfold to 519,192 t (ICLARM, 1998). Opportunities for further increases, especially in Bangladesh, the Philippines, and Viet Nam, have been enhanced by the genetically improved Nile tilapia strain from the International Center for Living Aquatic Resources Management (ICLARM), which reduces costs by 20–30 percent and which will make tilapia more affordable. Tilapia are herbivorous and, therefore, can be fed at low cost, and they are easy to breed. Moreover, preliminary research findings in Viet Nam also suggest that tilapia help to clean stagnant water in ponds and sewage areas (ICLARM, 1998).

The growth potential of Asia's inland fisheries and aquaculture is constrained by the limited availability of water of suitable quality, and inadequate property rights for land and water resources (Ruckes, 1996). It is also threatened by land-based pollution, and the environmental impact generated by the mismanagement of aquaculture (Box I.5). The introduction of new species and the transfer of species to new locations could also help spread diseases and pathogens, or endanger native species and thereby affect biodiversity. The explosive growth of aquaculture necessitates more public investment in R&D in fisheries. This is especially true because, despite fish being a major source of food security, R&D in fisheries has lagged behind that in food crops and livestock (Williams, 1996). As future growth and food security will have to come from culture rather than capture fisheries, more attention should also be given to research, development, and monitoring of genetically improved species and strains, both from indigenous and foreign sources.

FACTORS UNDERLYING GROWTH

Institutions

The State, the market, and the community all affect and can help coordinate resource allocation. The efficiency of the market is the result of free competition and the profit motive, which tend to force economic agents to maximize private benefits while minimizing costs. The State is more efficient in providing large-scale public goods, or in supplying goods and services in situations of decreasing cost under which monopolies tend to emerge, or when free riders and externalities (when costs are not borne by the actor, but by the public at large or by those not involved in the act, e.g. pollution) tend to be prevalent. In many instances when the boundaries of products are not defined, the State may create a market by establishing property rights.

The community is more efficient in providing and regulating local public goods and services, where social norms and values remain effective in regulating the behavior of local economic agents. The role and the timing of their interactions are important underlying factors in the success of Asian agriculture. The changing roles of these three actors in rural transformation are examined in more detail in a companion volume (Ammar, 1999). In this volume, only their current role is discussed, focusing mainly on the role of the State in agriculture.

The success of Asian agriculture is the result of interactions between these major entities and their relative influence at certain critical moments. The green revolution was launched at a time when the power of the State in most Asian countries was for the most part preeminent. Commercial opportunities were limited because most Asian farmers were subsistent or semi-subsistent producers. Local markets were small and only surplus output was traded. The lack of adequate transportation infrastructure isolated local communities from the central authority. Local communities were left to manage and mobilize local resources and set up the necessary arrangements for local public goods and services, such as small irrigation systems, fire and theft prevention, temples, and public ponds.

However, over the last two decades, the State has provided substantial infrastructure support, such as irrigation and road networks. During 1976 to 1995, road construction expanded by about 2 percent per year in the PRC. In India, the annual rate of road construction rose from 2 percent between 1976 and 1985 to 5.8 percent from 1986 to 1995. Empirical investigations into the impact of government expenditure on roads in India have indicated a highly positive impact both on agricultural growth and poverty reduction (Fan, Hazell, and Thorat, 1998).

The most significant contribution of the State lies in the delivery of and support for modern technologies. The green revolution's technology package was a publicly created product. Therefore, the role of technological extension was considered crucial to its delivery. A growth-accounting exercise conducted as part of this overall study (Rosegrant and Hazell, 1999), indicates a substantial contribution of public investment for research into productivity growth between 1972 and 1993. This investment occurred in most Asian countries except the Philippines and Myanmar. In addition to the delivery of technology, most countries actively engaged in creating infrastructure support, such as irrigation and transport systems.

The exogenous nature of the green-revolution package also demanded that the State reach out to farmers. The extension support provided by the World Bank known as the Training and Visit system further reinforced the government's capacity to deliver the package. The top-down or command-and-control nature of the public extension system complemented the delivery of the package during the initial stage. Many countries launched large public irrigation schemes to the make green-revolution technology viable. The State further encouraged its adoption by providing subsidized inputs and credits, and also implemented various schemes involving price supports and guarantees. In the 1960s and 1970s, concern over food security gave the government an excuse for widening its role in controlling the agricultural sector, often resulting in a plethora of government interventions. The tradition of top-down or centralized decision making has lingered on to this day, even though needs have changed.

Another very important function of the State is the provision of education. In Thailand, for example, education accounted for almost 40 percent of the increase in agricultural output between 1961 and 1985, exceeding the contributions of land expansion, irrigation, and increases in capital (Ammar et al., 1989). For the periods 1976–1978 and 1983–1985, the contribution from education to output was even higher, reaching 80 percent. One reason why education is such an important variable in Thailand is that as an Asian country with relatively abundant land, Thailand has only partially adopted green-revolution technology; the country has paid greater attention to the eating quality of its crops than to increasing yields.

In countries where free markets have been allowed to operate, the increased production encouraged trade and associated activities such as transportation, warehousing, and processing. Local markets started to proliferate, not only related to growth in output; input and credit markets also became widespread. In parallel with government machinery, such markets have become another important institution for resource allocation. Private businesses have emerged not only as suppliers of goods and services but also of technology production, adaptation, and transfer. In India, even markets for water have started because private pumping of irrigation water has become a profitable proposition.

The incentives provided by market and institutional reforms manifested themselves most vividly in countries where individual production decisions had been suppressed. The social and political forces that brought about these institutional reforms are treated in greater detail in a companion volume (Ammar, 1999). Decollectivization and market reforms in the PRC and Viet Nam greatly enhanced output. Towards the end of the 1970s, the PRC started implementing economic reforms and allowed market mechanisms to play a role in promoting agricultural production. The success of reforms in the PRC persuaded Viet Nam to follow suit.

In the PRC, rapid average annual growth in the agricultural sector between 1978 and 1984 of 7.7 percent, compared with an average of 2.9 percent during 1952 to 1978, was the result of a package of policy and institutional reforms consisting of price reform, the decollectivization of markets, and planning reform (Lin, 1992). Quota and nonquota prices were substantially raised. Collective land was reassigned to individual households under 15-year contracts. Decollectivization greatly improved the incentives for farmers. A policy of self-sufficiency was abandoned and imports of grain were allowed. The State no longer controlled crop choice. During 1978 to 1984, growth in total output in the PRC was 42 percent. Decollectivization alone accounted for about half of that growth. A subsequent study indicated that such institutional changes could provide added incentives to adopt innovations that would have long-term impact.

In Viet Nam, rice yields increased from 2 to 3.5 t/ha in just a decade, despite deteriorating irrigation systems and the lack of functioning extension and credit systems. Annual exports of rice increased from almost zero to over 2 million t in the early 1990s, despite the country's dilapidated agricultural infrastructure. By 1990, there was a tendency to recognize the market as an effective mechanism for resource allocation, even in Central Asia. As the food security situation improved over time owing to an increasing food production capacity and an improved capacity to earn foreign exchange, market distortions in favor of food crops no longer became necessary, although they were normally retained for social and political reasons.

Ownership of natural resources has tended to be claimed by the State on behalf of the public. Command-and-control regimes, supported by numerous pieces of legislation, are imposed on all natural resource sectors. Actually, most natural resources are de facto open access and free to all. The objectives of State intervention in the natural resource sector have been mainly related to production or revenue raising. Among the various natural resources, land was the first to be privatized; property rights were assigned and markets for land established. However, private property still accounts for only a small proportion of total land. For example, about 75 percent of Indonesia's land is under the control of the Ministry of Forestry (ADB, 1998). As population pressures increase, most governments have found it increasingly difficult to protect natural resources from encroachment, unscrupulous and illegal theft, and fires.

The State apparatus that was so essential and successful in the first decade of the green revolution became less effective with expanding local and international markets, with natural resources that were no longer abundant but degraded, and with a greater diversification and commercialization of agriculture. The State is also less useful when dealing with problems of food scarcity and security that are no longer widespread but limited to certain areas or regions only. The centralized commodity approach to which the State has become accustomed is not suited to the diverse conditions in rural areas that need to be brought under production.

While most States in Asia have accepted the potential of the free market, the same cannot be said about the State's perception of the potential of communities. Conflicts between the State and its people continue to deepen, especially in the competition for the use of public resources such as forests, land, and water. A few governments have now devolved limited responsibilities for managing public resources to local governments and communities. For example, community-owned irrigation systems, some of which have almost a thousand years of tradition behind them, have been legitimized by the State, and some of them have been linked to public irrigation schemes. Recently, new models of self-reliance developed by popular organizations, nongovernmental organizations (NGOs–e.g. Box 1.6), and local governments for the protection of natural resources have emerged at the community and watershed level.

**Box I.6   NGOs and Technology
           for Sustainable Agriculture**

In 1971, the Mindanao Baptist Rural Life Center, a nonprofit civil organization in the Philippines, initiated a project aimed at improving the livelihood of upland farming households. A technology package for sloping agricultural land (SALT) was developed and adapted by means of a participatory approach to suit the needs of farming communities. Inherent in the package were the following properties: (1) food priority, (2) production efficiency, (3) small farm focus, (4) reliance on internal resources (i.e. credit free), (5) economy of time and labor, and (6) environmental conservation and improved soil structure and fertility. The package had also to be culturally acceptable, economically feasible, and ecologically sound.

Elements in the package included contour preparation, the establishment of hedgerows, alternate strips, and hedgerow trimming for green manure. The technology was proven to be more economical than the terraced or the bench terrace technologies. It has been disseminated through the Baptist network and was discovered to be appropriate in areas with the dual problem of land degradation and food deficits. It was finally accepted as an official program by the Philippine Department of Agriculture in 1991, and has been revised, demonstrated, and adopted in many areas of the Philippines, Thailand, and Nepal.

Sources: Watson (1987); Serrano (1988); Tacio (1990).

It should be noted that the perceived need for a change in roles of the three entities cited above and the relationships between them have not always evolved smoothly and harmoniously. In many instances, lessons were taught and learned the hard way. At least one lesson is clear: the role of the State needs to change. The Asian experience (described in the next section) shows that an institution that is very appropriate for one task may be inappropriate when dealing with the same issue when surrounding conditions change. This is clearly shown to be the case in research, development, and extension. Visionary thinking and participatory planning and management are increasingly considered necessary in order to continue to adapt existing institutions to fit the changing environment.

Technology

As noted above, technology has played a crucial role in the phenomenal growth of Asian crop production. In most reports on the subject, technological changes in crop production have generally been related to green-revolution innovations: HYVs of rice, wheat, and maize, and increased use of inputs, particularly fertilizers and irrigation. However, as can be seen from the growth trends in the previous section, another momentous change that occurred during 1977 to 1997 and that contributed significantly to Asia's growth in crop production was the diversification of cropping systems. Thus, contributions to Asia's crop production growth came from two very different sets of innovations, "new" crop varieties and new management practices.

Crop Varieties

The plant genetic resources that have contributed to yield growth include modern varieties (MVs) and traditional varieties. Included in the MVs are high-yielding varieties (HYVs) that generally refer to the green-revolution rice and wheat that originated in the internationally-funded breeding programs spearheaded by IRRI and CIMMYT, and also hybrid rice (mainly in the PRC) and hybrid maize. These HYVs, combined with major increases in the use of fertilizers and irrigation, were the primary drivers behind the greater than 3 percent annual yield growth in these crops. By the early 1990s, 74 percent of Asia's wet-season rice area was planted with MVs (Hossain and Pingali, 1998), and dry-season rice crops are invariably planted using HYVs.

Countries vary a great deal in their adoption rates of MVs. In addition to the industrialized economies of Japan and the Republic of Korea, which grow MVs on all their riceland, those countries in which more than three quarters of their rice crop are planted with MVs are the PRC (100 percent, with 40 percent of the crop consisting of hybrid rice), Indonesia, Malaysia, the Philippines, Sri Lanka, and Viet Nam. Countries using MVs on one half to two thirds of their riceland are India, Myanmar, and Thailand. Countries planting only one third to one half of their riceland with MVs are Bangladesh, Nepal, and Pakistan. Bhutan (12 percent) and Lao PDR (2 percent) have the lowest adoption rates. The diffusion of wheat HYVs seems to have been faster than that of rice HYVs. For example, in 1991/92, HYVs accounted for 84 percent of India's wheat area, compared with 67 percent for rice. In 1994, MVs accounted for 91 percent of Asia's wheat area outside the PRC (up from 69 percent in 1977), and 70 percent within the PRC (Pingali and Rajaram, 1998).

The IR8 rice strain, developed by IRRI, was the first HYV to break the yield barrier in rice and started the green revolution outside the PRC. Less publicized was the fact that by 1965, the year before IR8 was released, 3.3 million ha of Guang-chai-ai, the PRC's own semidwarf rice, were being grown in the southern provinces of Guangdong, Jiangsu, Hunan, and Fujian. Guang-chai-ai is similar to IR8, and actually contains the same dwarfing gene as Dee-geo-woo-gen, one of IR8's parents (Barker, Herdt, and Rose, 1985). India also brought out its own semidwarf HYV, Jaya, in 1968. Since these first successes, improvement in the yield potential of rice has been limited.

Some 20–30 percent gain has been made by hybrid rice in the PRC; IRRI's new plant type holds promise of about 25 percent increase, but the new plants are still to reach farmers' fields. Hybrid rice has so far been restricted to the PRC because of the high cost of seed as well as the poor grain quality. Also, the average yield of PRC's hybrid rice has remained steady at 6.6 t/ha, the level it reached in 1986 (Yuan, 1994; Mao, 1994).

While the ceiling yield potential of rice derived from IR8 has not been overtaken by the succeeding HYVs, yield improvements in wheat and maize have continuously raised their ceiling potential. This partially explains their more robust yield growth trends.

The genetic yield potential of new wheat varieties continued to increase, by 0.5–1 percent per year, since the first semidwarf varieties came out of Mexico in 1964 until 1990 (Waddington et al., 1986; Byerlee, 1990). Breeders are optimistic about prospects for substantial increases in wheat yield potential over the next 10 years. Further incremental gains are being made from the important plant breeding innovation of wide crosses, which helps to create a "genetic bridge" between cultivated wheat and its wild relatives and allows researchers to draw upon the genetic wealth of those hardy species.

Crosses between elite varieties of durum wheat (containing the AB sets of wheat chromosomes) and goat grass (*Aegilops squerosa*), one of the original parents of bread wheat (containing the C set of wheat chromosomes), have given rise to a new "synthetic" wheat (with the complete three sets of ABC chromosomes of regular bread wheat). Through this synthetic but true bread wheat, desirable traits are being transferred more readily into elite bread wheats. This germplasm has been incorporated into CIMMYT's regular breeding programs, and the genetic material is already included in many advanced lines that have been tested in wheat-growing countries in Asia for several seasons now.

The growth of maize yields in the PRC during 1987 to 1997 of 2.76 percent per year was only half that of the preceding 10 years. However, Chinese hybrid maize breeders expect gains of about 10 percent in the yield potential from each generation of hybrids, i.e. 1 percent per year with about 10 years for each cycle (Chen, 1995). The capacity for yield growth is even greater for other Asian countries, where the diffusion of hybrid maize has only just begun. In less favorable areas there are also considerable yield gains that can be realized by the use of improved open-pollinated varieties, which have contributed significantly, even if not as spectacularly as the hybrids have done, to productivity gains in South and Southeast Asia over the last 20 years.

The diffusion of HYVs of rice, wheat, and maize is an on-going process in which the HYVs continue to be replaced by successive generations of newer varieties. The need to sustain growth so far won, especially against emerging pests and pathogens, has been as important as the push for further productivity gains and has been made possible by a very important change in Asia's agricultural research and development (R&D) institutions, i.e. the increasing capacity for plant breeding in the national agricultural research systems. The national breeding programs have not only helped to fine-tune varieties for specific agro-ecological niches, such as adapting to local tastes, market preferences, and soil and climatic conditions, but also have contributed some important productivity gains.

The other MVs include all the new varieties of rice and maize as well as of other crops (fibers, oils, roots, sugar, tea, coffee, tobacco, and rubber) that have also benefited from the innovations of modern plant breeding. In addition there are some traditional varieties of crops that have found new agro-ecological niches (e.g. basmati and other special quality rice varieties).

Although the yield growth of these other MVs and traditional varieties has been more moderate, they have contributed to Asia's crop production over the last 20 years in two ways. First, unlike the HYVs, which are restricted to irrigated areas, many of the less high-yielding varieties have been tailored to or found their place in less favorable areas. Examples of these include the open-pollinated maize for lower-input production systems and rice varieties, such as rainfed rice, deepwater rice, and upland rice, for difficult environments. Secondly, the nonHYVs were an essential part of the technology that between 1977 and 1997 drove production growth in crops such as oilseeds, vegetables, and fruits to rates comparable to or even higher than those of the green-revolution HYVs.

In order to meet the varying needs of specific locations, the All India Coordinated Research Project on Soybean released a total of 50 new varieties between 1982 and 1995 (Bhatnagar, 1995). While there was little evidence of any yield growth, the varieties were nevertheless an important instrument for the expansion of the soybean area in India, which grew from almost zero in 1977 to more than 5 million ha in 1997. Significantly, the biggest expansion of soybean farming has been into those states with the least irrigation, e.g. Madhya Pradesh, Rajasthan, and Maharashtra.

In spite of the current rapid diffusion of hybrid maize in India, driven by a strong private seed industry that was itself the product of a liberalization policy in the 1980s, a collaborative study between the Indian Agricultural Research Institute and CIMMYT has highlighted the fact that the hybrid maize revolution has hardly touched states such as Rajasthan, Uttar Pradesh, and Madhya Pradesh where maize is grown as a food crop and with a very low level of inputs (Morris, Singh, and Pal, 1998).

Thailand provides another example of the less high-yielding traditional varieties that have found new agro-ecological niches. These traditional varieties are characterized by favorable prices for the special quality of their grains and their ability to grow in physical environments that are unsuitable for HYVs. The better prices fetched by special quality varieties have helped rice farmers on saline/sodic soils (northeast region) with no irrigation and farmers in deepwater areas (central plain) to maintain the value of their crops in the absence of yield growth.

The spread of less high-yielding varieties has given rise to two issues that relate to the future of Asian productivity and its capacity for further growth. The first is the importance of local plant genetic resources, in terms of meeting specific agro-ecological conditions as well as being a source of germplasm for crop improvement. Access to this important source of future growth is threatened by current developments concerning the expansion of the scope of intellectual property rights to cover plant genetic resources.

The second issue is related to declines in the relative value of the HYVs. This will certainly mean slower growth or even a decline in the productivity growth of food grains, especially rice and wheat, in the future. Attempts to address this second issue have come from the national breeding programs that are now beginning to produce newer generations of MVs with special grain qualities. For example, the HYVs Pusa 4-1-11 for fine grain white rice, Annapurna and TKM 9 for red rice for the Kerala market, and Pusa Basmati for the aromatic basmati-type grain, have been developed in India. Thailand has also just released two new MVs, Fragrant Khong Luang 1 and Fragrant Suphanburi, neither of which is high yielding, but both are sufficiently insensitive to photoperiod to be grown in the dry season.

Crop Management

The crop management innovations that helped to increase crop productivity in Asia between 1977 and 1997 were of two main types: (a) an increase in use of inputs, especially fertilizers, irrigation, and pesticides; and (b) the increase in efficiency (increasing the efficiency of input use, and increasing cropping intensity, i.e. the number of crops grown in succession on the same land in one year). It should be recalled that although many of the inputs accompanied the green-revolution HYVs as part of the "new technology package", much of the increase in input use was also due to new high-value crops, especially fruits, vegetables, oil crops, and cotton.

Most of Asia's expansion in irrigated land took place in the 1960s and 1970s. Subsequently, the use of inputs increased at a much faster rate (Table I.9a). The rate of growth of fertilizer use and mechanization increased faster in the PRC than in Asia as a whole, but the use of pesticides in the PRC declined sharply

Table I.9a: Intensification of Input Use in Asian Crop Production, 1977–1996

	1977–1979	1994–1996	Increase (% per year)
Fertilizer use (kg/ha arable land)	60.18	133.55	8.3
Pesticide imports ($/ha arable land)[a]	1.669	3.628	7.5
Irrigation (% arable land)	30.50	35.75	1.0
Agricultural machines (units/1,000 ha arable land)			
Tractors	6.361	13.471	7.1
Harvesters/threshers	1.788	3.699	6.3
All machines	8.481	17.576	6.7

[a] from 1979

Source: FAOSTAT Database. *Available: http://apps.fao.org*

Table I.9b: Input-Use Trends in the PRC, 1977–1996

	1977–1979	1994–1996	Increase (% per year)
Fertilizer[a] use (kg/ha arable land)	57.33	176.70	12.3
Pesticides (kg/ha)	9.94	5.87	-3.3
Irrigation (% arable land)	30.2	33.0	0.5
Machines (billion watt/ ha)	0.79	2.28	11.7

[a] equivalent to ammonium sulfate with 20% N

Source: Adapted from Fan (1997).

(Table I.9b). The use of farm mechanization accelerated in the 1980s when rapid economic growth sharply raised the opportunity cost of labor, especially in Asia's Pacific-rim countries, and especially in the PRC and Thailand. However, owing to an existing large pool of labor in Bangladesh and Indonesia, the number of workers per tractor in those countries still remains high (Rosegrant and Hazell, 1999). The numbers in these two Tables should be used with caution as they do not reflect the size or the power of the tractors and machines.

The growth trends discussed in previous sections are clear evidence of the impact of input intensification. However, in addition to the obvious gains there have also been problems.

Pest Control

Crop losses from pests have always been a problem for Asian farmers. The availability of relatively cheap chemical pesticides since the Second World War has led to rapid increases in their use for "crop protection". The first pesticides used were mainly insecticides, and these have been applied to high-value crops such as vegetables, fruits, cotton, and plantation crops since the 1950s. Later, in the 1970s and 1980s, they were complemented by herbicides. The herbicides were used to lower production costs, either by reducing the cost of hand weeding or by enabling the use of certain labor-saving practices, such as broadcasting rice instead of transplanting it. Increases in horticultural production have led to the increased use of chemical pesticides, furthering the need for the greening of horticulture (Box I.7).

Box I.7 Greening the Production of Asian Vegetables

As incomes have risen in Asia so has the demand, and therefore supply, of vegetables. The major challenge to many developing countries is to produce quality products at affordable prices (Nangju, 1996). Vegetables typically require heavy investment in chemicals and are grown most profitably in highlands. This generates environmental impact in the form of soil erosion and chemical residues.

Research conducted at AVRDC and financed by the Asian Development Bank has resulted in high-yielding varieties of yard-long beans, tomatoes, hot peppers, and cucumbers, which have been distributed to farmers. Varieties that are resistant to disease have been identified. IPM and the use of a bio-insecticide *(Bacillus thuringiensis)* have been implemented. Varieties suitable for hot and humid regions have also been developed for the lowlands, thus spreading the benefits of high-value crops to the lowlands and easing the pressure for encroachment into fragile highland ecosystems.

The problem of pesticide use in foodgrain production is mostly associated with rice and is a consequence of the green revolution. In order to reduce crop losses from insect pests, the technology packages that delivered the first HYV (IR8) seed to farmers almost always included insecticides, usually one of the extremely potent organochlorines.

However, the organochlorines killed not only the insect pests but also their natural predators. Insect ecologists tried to draw attention to this from the early 1960s, but were ignored for the most part. Then the insect pests began to develop resistance to the pesticides, especially to some of the organophosphates that were replacing organochlorines. Attempts to combat these developments proceeded by either increasing the dose or combining several chemicals into even more lethal pesticide "cocktails". These only worsened the situation because they served to kill even more of the pests' natural predators and further increased the evolutionary pressure on pests to develop even greater resistance to the pesticides.

The most severe consequences of these problems were brown planthopper epidemics. Previously, this insect had been an inconsequential inhabitant of Asian rice fields. However, with its natural predators being greatly diminished it became a menace to Asian rice crops, and the severity of this menace increased in direct proportion to the intensity of insecticide used. In northern Sumatra, Indonesia, farmers were treating their fields with pesticides 6 to 20 times over periods of 4 to 8 weeks, with no success (Kenmore, 1991, cited by Conway, 1997). The density of the insect pest population increased with the increasing frequency of spraying.

In addition to insecticides, herbicide use has also been increasing rapidly in Asia. By 1996, the value of herbicide imports was two thirds that of insecticides. No data are available on the efficiency or impact of herbicide use. Weed resistance to herbicides, which is now one of the very serious problems in crop production in developed countries, could also become a threat in Asia in the near future. Resistance to some of the most extensively used herbicides in Asia, such as isoproturon and propanil, has begun to appear.

The green-revolution technology itself has intensified the pest problem and in many ways has stimulated the increased use of pesticides. Large monocultures and the year-round planting of single crops create ideal conditions for massive pest outbreaks. The high levels of nitrogen in the applied fertilizers make plants more susceptible to certain pathogens (e.g. blight in rice) and insects.

Large government subsidies for pesticides have also provided a critical boost in pesticide adoption. In Indonesia before 1986, for example, farmers were only paying about 15 percent of the actual cost for pesticides. As a result, 20 percent of all of the pesticide applied to rice worldwide was being applied to Indonesian rice crops, although Indonesia accounted for less than 9 percent of total world rice production. In Thailand, there was no government subsidy, but government pest control units distributed pesticides free of charge when "outbreaks" were reported. Unfortunately, the term outbreak was loosely defined, and very rarely were the pesticide handouts economically or ecologically justified.

Fertilizer Use

After three decades of annual growth in fertilizer use of around 10 percent, Asia's croplands in 1996 received on average around 135 kg/ha of fertilizer, from 38 kg/ha in 1965. In irrigated areas, fertilizer consumption per hectare is much higher than the national average. Malaysia, PRC, Republic of Korea, and Viet Nam use over 700 kg/ha of fertilizer on irrigated land (Table I.10). In Indonesia, the rate is over 500 kg/ha. Countries with low consumption levels, such as Cambodia and Myanmar, tend to be those with foreign exchange problems, lack of proper distribution and credit systems, and lack of incentives resulting from price ratios of grains to fertilizers. Much higher rates are applied in intensive cropping systems. Up to 1,000–2,000 kg/ha of fertilizer are used in intensive vegetable growing in countries throughout the region (Morris, 1997).

Indications of inefficiencies in fertilizer use are much less obvious than in insecticide use. No major ecological or economic

Table I.10: Total Fertilizer Consumption in Irrigated Areas in Asia

	Total Fertilizer Consumption (kg/ha)			Annual Growth Rate (%)	
	1975	1985	1995	1975–1985	1985–1995
East Asia					
China, People's Rep. of	160.17	378.00	713.67	9	6
Japan	568.05	689.02	609.26	2	-1
Korea, Rep. of	677.76	609.06	714.73	-1	2
Mongolia	152.17	310.00	31.25	7	-23
Southeast Asia					
Cambodia	1.12	0.00	58.38	-1	41
Indonesia	125.41	458.56	558.17	13	2
Lao PDR	2.50	16.81	34.97	19	7
Malaysia	805.26	1,830.54	3,323.53	8	6
Myanmar	56.07	178.88	109.59	12	-5
Philippines	218.17	196.65	381.72	-1	7
Thailand	74.47	113.44	311.64	4	10
Viet Nam	330.00	217.85	724.00	-4	12
South Asia					
Afghanistan	14.97	28.21	17.86	6	-5
Bangladesh	149.49	260.82	372.59	6	4
Bhutan	4.55	3.33	2.56	-3	-3
India	103.58	203.55	276.97	7	3
Nepal	53.30	57.12	105.90	1	6
Pakistan	40.63	95.88	145.80	9	4
Sri Lanka	150.83	335.31	363.38	8	1

0 = zero or less than half of the unit measured.

Note: Annual Growth Rate = ((Ln(value year 1) - Ln(value year 2)) / number of years) x 100.

Source: FAOSTAT Database. *Available: http://apps.fao.org*

disasters on the scale of the brown planthopper epidemics have yet been reported. Many authors have, nevertheless, pointed to two possible types of impact of fertilizer use in terms of (a) fertilizer-use efficiency and (b) nutrient imbalances.

Declines in the ratio of grain to fertilizer, e.g. in India from about 60:1 in 1966 to less than 10:1 in 1992 for rice and from 15:1 to 5:1 for wheat, have caused concern about a possible decline in the efficiency of fertilizer use. This has, however, been indicated as being somewhat misleading because the grain:fertilizer ratio is not a very accurate indicator of fertilizer-use efficiency; it erroneously assumes a zero yield in the absence of fertilizer use (Hobbs and Morris, 1996). In Karnal (in Haryana, the heart of India's green-revolution territory), the marginal response to fertilizer (the ratio of increase in yield to increase in fertilizer) for rice has declined somewhat. This can probably be explained by the diminishing returns from the very high rates of fertilizer now being applied. For wheat in India, the marginal response was still increasing slightly in the early 1990s (Chaudhary and Harrington, 1993).

It is now technically quite simple to achieve the twin goals of improving fertilizer-use efficiency and soil nutrient levels. Losses from nitrogen fertilizers can be effectively minimized through the use of such innovations as urea supergranules or deep placement of urea, and urease inhibitors. Fertilizers that are well balanced in relation to crop requirements and the soil's own nutrient capacity can be easily formulated with help from soil and plant analysis and fertilizer trials. Most of these, however, are still not yet applicable for use on the average Asian farm. Most farmers consider the deep placement of urea as too labor intensive. Supergranules are very costly. The urease inhibitors, which were only in the experimental stage in the early 1990s, have not yet been incorporated into the fertilizer manufacturing process. Access to services that would help to improve the match between the nutritional content of fertilizer, the capacity of the soil to supply nutrients, and the needs of crops is still unavailable to most Asian farmers.

In developed countries, individual farmers sometimes use tools such as plant and soil analysis and fertilizer trials for fertilizer management. More often, however, these facilities are provided as part of the service rendered by fertilizer companies and farm consultants. Such services are rare in Asia. Among the exceptions are the consultancy services that provide advice to the larger oil palm and rubber plantations in Malaysia and Indonesia. Their fertilizer recommendations are generally based on tissue analysis. Some fertilizer companies that provide soil analysis services as part of their marketing operations are now found in the region.

Lack of analytical facilities is not the main reason for the lack of services to support improved fertilizer management in Asia. Most analytical laboratories, many exceptionally well equipped by development assistance programs, are actually greatly underutilized. Logistical arrangements are lacking on how to take samples, determining where to send them, making sure of the timely return of results, and interpreting the results. The services provided by the analytical laboratories are, therefore, of little use to district farm advisors, farmers, or fertilizer marketing personnel. Most analytical chemists would also point out that the results from plant and soil analysis are useless without stringent quality control on laboratory procedures.

Most farmers in Asia have few resources for fertilizer management other than the "official recommendations". These recommendations are not generally very responsive to local variations or the impact of cropping intensification that has taken place in the last 20 to 30 years. For example, despite the thousands of fertilizer experiments that have been conducted in the past two decades throughout India and Pakistan, practically the same fertilizer recommendation is given in all irrigated areas (Byerlee, 1990).

Some experts are now advocating alternative agriculture, defined as alternatives to high-input technology such as that of the green revolution (Box I.8). However, more studies on costs and trade-offs are needed; there is no single formula that fits all circumstances.

Breeding of efficient varieties offers one widely adaptable solution, at least for deficiencies in some of the micronutrients. Such a strategy has been shown to be highly feasible in dealing with boron deficiency, which causes widespread yield losses in wheat in the southwestern PRC, Bangladesh, northeastern India, and Nepal. Boron deficiency causes the local standard varieties to set grain poorly. Many boron-efficient genotypes have been identified among CIMMYT's advanced materials (lines that have been widely tested and are almost ready for release as varieties), which will set grain normally. Simple screening has prevented inefficient varieties from being released into problem areas in Nepal. For much of the wheat-growing area of Bangladesh and the northeastern states of India, such as Bihar, Orissa, and West Bengal, it would surely be better to screen inefficient germplasm out before it is released, only to be rejected by farmers after it fails to set grain properly, as is currently the case.

Box I.8   Organic Farms

Organic farms have attracted the attention of many NGOs, and are sometimes held up as a general solution for agriculture. Organically grown or chemical-free agricultural products have now found a niche in high-income markets. Although these markets are small, they are expanding. This market niche, which attracts premium prices, is important for organic farm products because the cost per unit of organically grown produce is often higher than chemically grown alternatives (NRC, 1989). In Asia, there is little definitive evidence of this cost difference (i.e. when the alternatives are compared in terms of unit cost and net natural resources consumed). Most anecdotal evidence of success consists of production method and yearly income only. However, it has been observed that the more successful cases are those involving expert farmers; a high level of expertise is required.

The basic principle of organic farming is an emphasis on green manure and nutrient recycling. This principle has some trade-offs and limitations of its own. For example, rice fields fertilized with green manure legumes are likely to release much more methane than those fertilized with urea or ammonium sulfate; also, as greenhouse gases, nitrogen oxides from fertilizers are dwarfed by methane in quantity. In other words, nitrogen fertilizer is greener than green manure in this respect. Grain legumes may save the cost of nitrogen fertilizer, but they quickly run the soil into nutritional imbalances as they can deplete the soil of many other nutrients, e.g. potassium and calcium, much more quickly.

Organic farming can be an attractive alternative for farming in relatively favorable areas. Most prototypes of organic farms are in areas where soil fertility is favorable. However, in areas where a deficiency of nutrients other than nitrogen is a major problem, organic farming methods cannot provide nutrients to recycle. For example, in some parts of the Lao PDR, the soil is so deficient in phosphorus that cattle

(continued next page)

Box I.8 (continued)

manure does not offer a solution for improving soil fertility. Furthermore, cattle that are deficient in phosphorus do not reproduce; there is a cycle of not enough cattle, not enough manure, and, of course, little phosphorus in the manure. On some of the acidic soils of the mountainous areas of Viet Nam, legumes do not fix much nitrogen in the soil because of either acidity or molybdenum deficiency. Plants need 15–16 macro- and micronutrients; organic farming is not viable if there are no nutrients to be recycled.

Research, Development, and Extension

In the early period of the green revolution, 1971–1980, productivity growth accounted for about one third of the total growth in Indian rice production. This productivity growth was related to R&D, canal irrigation, the balanced use of fertilizers, and agricultural terms of trade (Kumar and Rosegrant, 1994). Public research alone accounted for about half of the productivity growth. Investment in technology is indisputably an indispensable input for sustainable development.

Four important factors in agricultural R&D institutions in Asia are (i) growth of national plant-breeding capacity, (ii) lagging capacity for generating crop management innovations, (iii) inadequate public extension systems, and (iv) evolving roles of the private and NGO sector in agricultural R&D. These have been instrumental in the technological changes discussed above and will be crucial in determining the future capacity for growth and sustainability of Asian agriculture (see also Rosegrant and Hazell, 1999).

National Plant-Breeding Capacity

The capacity for plant breeding in Asia has grown enormously in the last 20 years. While the first rice and wheat HYVs came to Asia directly from IRRI and CIMMYT, contributions to later generations of HYVs have come largely from national plant breeding programs. The increasingly active national agricultural research systems (NARS) have not only helped to fine-tune varieties for specific agro-ecological niches, but have also contributed some important productivity gains. Many countries in Asia now appear confident of their own rice-breeding capacity. A recent consultation, organized by IRRI, of national agricultural research leaders from Bangladesh, PRC, India, Republic of Korea, Philippines, Thailand, and Viet Nam reached a consensus that conventional plant breeding should be the responsibility of NARS and not IRRI.

For Asia's main crops–rice, wheat, maize, soybean, groundnuts, and mungbean–all the national breeding programs have drawn heavily from the germplasm support services of the CGIAR centers and AVRDC, which provides germplasm for soybean and mungbean. The important germplasm services provided include conservation, enrichment (in which new genes are introduced through new plant-breeding innovations such as intraspecific and other wide crosses as well as genetic engineering), and the transfer of genetic materials. This multilateral arrangement for germplasm management is unmatched by any other facility in its cost effectiveness and equitable sharing of plant genetic resources.

National funding of germplasm conservation tends to be very limited (NRC, 1991). Without multilateral arrangements for germplasm exchange, individual country access to plant genetic resources would be extremely limited. The international public germplasm services provided by the CGIAR centers and AVRDC will continue to be essential for the maintenance of the genetic diversity of Asia's (and the world's) important crops as well as for the capacity of national breeding programs to meet the need for long-term sustainable crop production. These services have, however, now been put under threat by a major controversy relating to the expansion in scope of intellectual property rights to cover plant genetic resources (More details on this are provided in a companion volume (Ammar, 1999)).

Capability for Generating Crop Management Innovations, a Critical Weakness

The green-revolution technology of MVs was widely adopted, and the diffusion process was relatively simple. Many of the problems that have since developed are much more location specific. The crop management innovations required to solve these problems as well as to provide further productivity gains need to be sensitive to the set of socioeconomic and biophysical conditions particular to each location.

Asia's capacity for crop management R&D, however, lags far behind its plant breeding capacity. The situation in South Asia as described by Byerlee (1990) has relevance for the whole region: "The strength of plant breeding research in South Asia contrasts with the relative weakness of crop and resource management (that is research on tillage, fertilization, pest control, irrigation scheduling, planting date and establishment, and so forth)". He argued that the current system of largely centralized agricultural R&D, although managing well enough for the development of new MVs, is poorly equipped to generate and transfer effectively the much needed innovations in crop management.

The need for crop management R&D to focus on the agro-ecological and socioeconomic differences between farmers was recognized by the early 1970s, when the green revolution was just beginning in Asia. That the need still remains today speaks clearly of the failure of attempts to deal with the problem in the intervening years. Notable among these was the farming systems movement that became fashionable and absorbed enormous amounts of resources in the 1970s and 1980s.

The lack of progress after such a long time prompts the conclusion that perhaps the benefits that can be expected from crop management research are limited. However, in South Asia, major productivity gains have been realized from research in crop management on problems such as the date of sowing, crop establishment, and weed control in the rice-wheat cropping system (Hobbs, Sayre, and Ortiz-Monsterio, 1998).

The PRC appears to be best meeting the need for crop management innovations. An important reason for this is the system of agricultural R&D, which is to a large extent locally controlled. The research extension network, involving the county, commune, brigade, and production team, provides a mechanism for rapid evaluation and selection of hybrids for local adoption and diffusion of technical information related to their management.

These results also illustrate an important weakness preventing effective transfer of crop management innovations, and which is why they are much less widely adapted than are the HYVs: crop management R&D in most of Asia is largely supply driven and not a response to farmer demand, making farmers less willing to adopt it. The following remark referring to Indonesia makes the point: "Researchers are oriented toward the publication of findings in ministry journals and magazines, not toward solving problems. The reward system (promotion, incentives) is structured that way" (Manwan et al., 1998). Manwan's remark is also applicable in many other countries throughout the region. There is a critical lack of feedback mechanisms between the users (i.e. farmers) and the producers (agronomists, research scientists) of crop management innovations. Many countries in Asia have made great efforts to decentralize their agricultural R&D process. Unfortunately, the results have mostly been simply the creation of yet another layer of bureaucracy. The two elements that are important to the generation of effective crop management innovations for specific agro-ecological niches are (a) the flow of information from farmers on problems that need solving, and (b) the capacity to respond effectively. Unfortunately, these are still largely missing in most Asian countries.

In the PRC, an essential measure of quality control on crop management innovations has been provided by a contract system in which local agricultural officers share with farmers the rewards of production increases as well as punishment for failures incurred by any innovations they have suggested. The capacity for crop management R&D in the PRC can be expected to be much enhanced by the improved capacity in agricultural science that has taken place in the last 20 years. For example, obtaining an 18 percent yield increase (almost the gain made by the PRC's famous hybrid-rice technology) though improvements in plant nutrient management (Lin and Shen, 1994) seems highly feasible, given the more than 20,000-strong cadre of plant nutrition specialists in the PRC. Other nutritional limits to yields have also been identified in the new intensive cropping systems, e.g. widespread boron deficiency in rapeseed in Hubei and Zhejiang provinces (Lu et al., 1997; Wei et al., 1998).

In other parts of Asia, inadequate capacity in crop management, unless corrected, will continue to exact costs in three ways: (a) through yield and profitability losses, (b) by placing the system's sustainability under threat, and (c) through impact on human health and the environment.

A major obstacle to the decentralization of agricultural R&D and to farmers' participation in identifying R&D needs is the perception, still common among researchers and policymakers in the region, that the capacity to use complex technologies efficiently is limited by the low level of formal education among farmers in Asia. This has been proven wrong, not only for the PRC (some would argue that Chinese farmers are very different from South or Southeast Asian farmers), but also on a very large scale. One example is the IPM experience in Indonesia, where the transfer of crucial knowledge on insect ecology has been successfully done with farmers who have had no or very little schooling.

Public Extension Systems

In spite of the heavy international and national investment in farming systems research and extension in the 1970s and 1980s, agricultural technology transfer in Asia has remained very much "top-down". The approach has not worked well for crop management because there was insufficient consideration of onfarm conditions during the research phase, and inability to respond to second- and third-generation problems that emerged.

For example, agronomists and farmers in South Asia have known for a long time that delaying the sowing of wheat beyond the optimum planting date would lead to a yield loss of 1 percent per day. The simple solution of sowing earlier, however, was not an option for a large number of farmers in India and Pakistan, whose wheat crop is only one component in such systems as rice-wheat, cotton-wheat, and soybean-wheat. There were various legitimate reasons why the first crop could not be harvested early enough for the wheat to be sown by the optimal date. Reduced tillage was suggested as a solution because it reduces the turn-around time between crops. This appears to be a widely effective solution, but it has brought on another serious problem, herbicide resistance in some major weeds in the rice-wheat cropping system, because reduced tillage requires increased use of herbicides. Overcoming the problem will require feedback from the fields and further research effort.

As indicated earlier, except for the PRC, yields from farms lag far behind those from experimental stations. Productivity gains and environmental benefits can be achieved by plant nutritional balance and better water and soil management. This type of innovation requires two-way communication between scientists and extension officers on the one hand, and with farmers on the other. Such communication is not common in most research and extension activities in Asian countries. Most public extension systems have been criticized as being too centralized. In those countries where devolution has begun to taken place, local organizations lack adequate funding and lack linkages with R&D systems. Extension personnel tend to lack technical skills in crops other than rice, lack the flexibility and skills to adapt generic solutions to specific locations, have inadequate and irrelevant information vis à vis local needs, and are not responsive to farmers' interests.

As cropping systems have become mixed and complex, farmers have increasingly turned to other sources of technical information, e.g. private traders, factories, NGOs, and local universities (Mingsarn, Kanok, and Chaiwat, 1989; World Bank, 1996). However, farmers in unfavorable areas tend to have little access to public extension systems (World Bank, 1996). Further, public extension systems are generally geared towards male farmers. Female farmers tend to take care of subsistence crops or livestock, but have relatively less access to information on agriculture, including that concerning nutritional maintenance of, for example, livestock. Other means of extension, such as television, that reach women in their homes and overcome cultural obstacles limiting contact with the outside world, have rarely been used.

Evolving Roles of Private-sector and NGO Agricultural Research, Development, and Extension (RD&E)

The green revolution has basically been the product of massive public investment in agriculture. For the new generation of HYVs, there have been subsidies for inputs and the development of huge irrigation schemes. This level of public investment in agriculture is unlikely to be sustainable. Declines had already started in Asia by 1980 (Rosegrant and Pingali, 1991). Countries have been phasing out subsidies for various inputs, which, to a certain extent, may be desirable, as inefficiencies are often eliminated in the process.

The evolution of private-sector involvement in RD&E is more discernable in the seed industry. The liberalization of the seed industry in India since the 1980s has had a tremendous impact on the production of hybrid maize seed, which rose from almost none in 1984 to more than 12,000 t in 1992 (Morris, Singh, and Pal, 1998). Some 45 percent of India's maize area was planted with improved open-pollinated varieties and hybrids in the 1994/95 season. A production increase of more than one million t for that crop season resulted from the adoption of maize hybrids. Similarly, the process now taking place in the PRC, with support from the World Bank, is expected to transform the seed industry there. Private vegetable seed producers have also been active in other Asian countries.

Many have argued, backed by a strong lobby from the seed industry, that the protection provided by intellectual property rights (IPR) legislation and exclusive marketing rights to seed varieties should (a) provide incentives for private R&D in plant breeding, (b) save the public the cost of seed production and dissemination, and (c) earn revenue for public breeding programs through the licensing of varieties. There are probably some significant savings to be gained from privatizing seed production and dissemination. There may also be greater public benefits from more efficient seed production and dissemination through variety licensing than the return on revenues from royalties from breeding programs.

For hybrid maize, however, private investment has preceded IPR laws. In Thailand, where a proposed Plant Variety Protection Act had not been enacted as of 1998, hybrid maize operations have been carried out by national and multinational seed companies since 1979 (Pongsroypech, 1994). These operations have involved every step of the business, including the introduction of germplasm and breeding. By 1981, hybrid seeds, largely imported, were put on the market by several companies. It was not until 1990, presumably when the yield potential from hybrids was finally realized by more locally adapted materials, that diffusion began to accelerate.

Experiences from Argentina and Chile have shown that IPR for plant breeders has had very little impact on private R&D investment in plant breeding. This is due, in part, to limited enforcement mechanisms, to court procedures that have yet to be established, and to seed companies' unwillingness to take violators to court (Frisvold and Condon, 1998). In the US, it has been concluded that a "farmer exemption" (a clause in the IPR law that allows farmers to keep progenies, i.e. succeeding generations of seeds of "proprietary" seed, for their own use) was the main reason why most of the seed companies ceased wheat R&D, and perhaps also soybean breeding (Pray, 1991).

NGOs and the private sector have recently taken a very active role in technology transfer in crop management. NGOs in particular have been able to adapt appropriate site-specific technology to the needs of local communities (see Box I.6). In many instances, they help to combine and enhance traditional knowledge with modern technology. However, sometimes technology proven successful by NGOs has failed to be effective when handed over to government agencies. At that stage, the participatory approach may become substituted by a top-down, seedling- and fertilizer-subsidy mentality.

In conclusion, a constant challenge for international crop breeders is to continue to increase yield potential in order to keep production ahead of population growth. Past successes have emphasized land-saving technology. Additional requirements for the next decade are to include in the technology package elements that provide environmental savings and are environment enhancing. The technologies should not be limited to water and soil conservation but should also include development of varieties for more extreme environments, such as saline or acidic soils, and varieties that are mineral efficient. This type of R&D requires substantial collaboration not only with national breeding programs but also with local universities, researchers, and the farmers who possess first-hand knowledge of local conditions.

For the national R&D programs, the challenge will be in crop management innovation. These innovations may deal with 1) fine tuning new varieties to suit specific agro-ecological niches, 2) improving efficiency in input use, and 3) increasing cropping intensity coupled with more effective environmental management. Again, the challenge is to reverse the "father-knows-best" approach currently employed by RD&E workers to a "farmer-first" system. A greater challenge is for the centralized agricultural agencies in many countries in Asia to recognize traditional knowledge and blend it with science-based technology and management systems. Room for this exists in water resources management, local nonchemical herbicides and pesticides, and cropping for local conditions.

Moreover, breeding new crop varieties depends on a fair and transparent system of international gene-pool management and exchange of genetic material. It is evident that there is an urgent need for proper multilateral germplasm management that recognizes the rights of prior users and providers of genetic materials, as well as the need to create sufficient but not undeserved incentives for private R&D.

Irrigation

Irrigation was key to the success of the green revolution. Irrigation not only augments the water supply but also improves and ensures the stability of water delivery, widens crop choices, and allows increased cropping intensity. Asia has 179 million ha or 69 percent (in 1995) of the world's irrigated areas. The PRC alone has 49 million ha under irrigation, India 50 million ha, and Pakistan 17.2 million ha (Table 1.11).

Over the past two decades, irrigated areas have increased in most Asian countries, especially Bangladesh, Bhutan, PRC, India, Pakistan, Thailand, and Viet Nam (Table I.11), but the rate of increase has slowed down in the last decade except in PRC, Myanmar, and Bangladesh. Since 1980, the irrigated area in Asia has expanded at the rate of about 2 percent per year; about 35 percent of the arable land in the region is now under irrigation. Future growth in irrigated area may come from India where it is planned to add 17.3 million ha of irrigated land by 2020 (Rosegrant and Ringler, 1998). The general reduction in growth of irrigation is a result of the decline in funding by major lending agencies as well as of the difficulties in finding projects with high returns. In many countries, such as PRC, Japan, Republic of Korea, and Sri Lanka, the supply of land that would yield a high return under irrigation has been mostly exhausted. In other countries, such as India and Thailand, where expansion is being planned, the marginal cost of irrigation is high when social and environmental costs are taken into account.

Since 1980, the efficiency of irrigation systems has become an issue of increasing concern. Poor maintenance and rapid deterioration are common features of irrigation systems in many Asian countries. Irrigation agencies are interested in increasing physical capacity without commensurate increases in management capacity. Planned capacities have fallen short of

Table I.11: Proportion of Arable Croplands Under Irrigation in Selected Asian Economies

	Irrigated Area (ha'000)			Proportion of Arable Lands Under Irrigation (percent)			Annual Growth Rate of Irrigated Areas (percent)	
	1975	1985	1995	1975	1985	1995	1976–1985	1986–1995
Asia	121,165	140,792	179,013	27	29	35	1.50	2.40
East Asia								
China, People's Rep. of	42,776	44,581	49,857	43	46	52	0.41	1.12
Japan	3,171	2,952	2,700	62	62	62	-0.72	-0.89
Korea, Rep. of	1,277	1,325	1,335	57	62	67	0.37	0.08
Mongolia	23	60	80	3	4	6	9.59	2.88
Southeast Asia								
Cambodia	89	130	173	5	6	5	3.79	2.86
Indonesia	3,900	4,300	4,580	15	16	15	0.98	0.63
Lao PDR	40	119	177	6	14	20	10.90	3.97
Malaysia	308	334	340	7	6	4	0.81	0.18
Myanmar	976	1,085	1,555	10	11	15	1.06	3.60
Philippines	1,040	1,440	1,580	14	16	17	3.25	0.93
Singapore				0	0	0	0	0
Thailand	2,419	3,822	5,004	15	19	24	4.57	2.69
Viet Nam	1,000	1,770	2,000	16	28	30	5.71	1.22
South Asia								
Afghanistan	2,430	2,586	2,800	30	32	35	0.62	0.80
Bangladesh	1,441	2,073	3,200	16	23	37	3.64	4.34
Bhutan	22	30	39	20	23	26	3.10	2.62
India	33,730	41,779	50,100	20	25	30	2.14	1.82
Nepal	230	760	885	10	33	30	11.95	1.52
Pakistan	13,630	15,760	17,200	69	76	80	1.45	0.87
Sri Lanka	480	583	550	25	31	29	1.94	-0.58

0 = zero or less than half of the unit measured.
Note: 1. Annual Growth Rate = ((Ln(value year begin) - Ln(value year end)) / number of years) x 100.
Source: FAOSTAT Database. *Available: http://apps.fao.org*

actual needs, and some systems are unused owing to lack of water, inappropriate design, or poor maintenance (IRRI, 1983; Kikuchi, 1996). The overall system efficiency is low, for example 30 and 38 percent in northern India and Karnataka, respectively (Guerra et al., 1998).

In Central Asia, the breakdown of the drainage system in salt-affected irrigated areas has led to further elevation of the groundwater table, thereby increasing salinity, which has led to large yield losses and finally to the total loss of cropland in some areas.

Communal irrigation systems that have been taken over by centralized irrigation agencies often become inefficient because of a lack of appreciation of onfarm water needs. A study of 15 irrigation systems in South and Southeast Asia indicated the lack of two-way communication between irrigation agencies and water users (Murray-Rust and Snellen, 1993). "Flood" irrigation, which is the prevalent system in Asia, is in itself an important potential source of inefficiency and degradation. Prolonged or excessive flooding results in waterlogging and salinization. At the farm level, farmers tend to use more water than is needed. For rice, the amount of water used may be 6 to 10 times more than is necessary (Ghani et al., 1998). Water pricing has been suggested as a means of overcoming water waste, but farmers will then have to weigh the cost of water with the cost of weed control. Another method for saving water is the conversion from transplanting to direct seeding which reduces water use by half, although the yield may be lower even with good weed control. Other irrigation techniques, e.g. drip irrigation and methods that are site-specific water applications, are emerging where water is scarce and the crops are of high value.

In the past, the construction of multipurpose or agricultural large-scale dams was often planned top-down, with insufficient consideration given to the people who would be affected by the project. The impact on forests resources and biodiversity was usually not taken into account. Recently, some well-organized NGO networks have effectively publicized the plight of dam refugees and the ecological costs of large-scale infrastructure projects. This has rendered projects in the region more transparent and has enhanced the accountability and worthiness of the projects. In Thailand for example, a careful review of the feasibility of the Kaeng Sua Ten Dam in northern Thailand following protests by environmentalist groups revealed that the project proposed by the Government was not economically viable (TDRI, 1997). The rate of return was low even before the mitigation and environmental costs were taken into account.

The process for determining economic feasibility and environmental impact needs to be strengthened to improve irrigation efficiency and to avoid having dams that have insufficient inflows or that lead to an increase in soil salinity. To date, the feasibility study and the environmental impact assessment have been important only in terms of the loan application process. Irrigation projects are often not transparent or accountable to the public. Cost-benefit analyses have not been rigorously conducted. It is important that in the future, projects should involve the participation of a wider spectrum of stakeholders.

Recently, there has been global recognition of the value of consulting and involving water users in water management plans and activities related to irrigation systems. For the past two decades, more and more countries around the world have been turning over management authority for irrigation systems to farmers' groups or local entities, in a process commonly referred to as irrigation management transfer (IMT). There have been several studies on this process and the literature shows a mixture of positive and negative results (Vermillion, 1997).

Although most of the studies are deficient in assessing the real cost of farmers' participation, government expenditures for irrigation tend to decline and costs to farmers often rise. There is little evidence to suggest that yield, water productivity, or farm income has increased. Poor operation and management have a negligible impact on the irrigated crop. Studies that would make it possible to separate the impact of IMT from other factors such as weather are lacking. In many instances, the responsibility for rehabilitation is not clearly spelled out in the IMT agreement between the government and local entities.

The key to sustained success of farmers' participation is the incentive structure and quality of leadership, which can vary widely from place to place and from time to time. There is no available model to follow for molding the farmer-agency relationship that will work in all societies in all situations. Many innovations may be needed to develop the right relationship for a given set of conditions. It is hoped that as the real value of water becomes better understood by all users and as more realistic water pricing becomes feasible, workable models will emerge for sharing responsibility between agencies and users in managing irrigation water .

Urbanization

Urbanization, as measured by the percentage of the total population living in urban areas, has been increasing steadily in Asia for decades. The average annual urbanization rate for the whole of Asia between 1980 and 1985 was 3.6 percent (WRI, 1998) and the predicted averages for 2000–2005 and 2020–2025 are 2.8 and 2.0 percent, respectively. Urbanization occurred especially rapidly in parts of East and Southeast Asia. Nevertheless, the region is not yet highly urbanized, with about 30 percent of its inhabitants living in urban areas as of 1990. When compared to urbanization in other developing regions, Asia is close to Africa where only about a third of the population lives in urban areas.

Internal migration has fuelled much of Asia's urbanization, with approximately 60 percent of urban growth coming from rural migration. Economic growth has usually been accompanied by declining fertility rates. Therefore, the contribution of natural growth to urban population increases is likely to decline relative to that of rural migration as Asia's economies continue to develop. This can be seen in the markedly slower growth of rural populations. Moreover, as younger and more educated persons leave rural areas, farms are being managed by older generations and by women. This has happened in Japan and the Republic of Korea, and more recently in the PRC, Malaysia, and Thailand.

Rapid urbanization and industrialization in Asia have both positive and potentially negative effects on agriculture. Increased wages in urban and industrial areas lure labor away from agriculture. For high density areas, this rural-to-urban migration raises productivity in agriculture, but for a country such as Thailand where land is abundant, rapid growth has increased the cost of labor dramatically (Coxhead and Jiraporn, 1998), necessitating substitution of capital for labor.

In the past decade, Asia has gone through a period of rapid economic transformation. Urbanization and industrialization have taken land from agricultural production. Fertile and irrigated areas have been converted into housing estates and factories. Expanding urban markets often demand greater amounts of horticultural and livestock products, which lowers incentives for the production of grain. The annual loss of wet riceland to urbanization in Indonesia has been estimated by the Agency for Agricultural Research and Development at between about 20,000 and 100,000 ha per year. There have been fears that the PRC is on the verge of not being able to feed itself, based on a large expected loss of cropland in the next few decades as a result of industrialization and urbanization (Brown, 1995). However, Lindert (1996a) estimated that agricultural land lost to urbanization in the PRC has been quite small, i.e. 0.04 percent per year between 1983 and 1993. In addition, other studies (Wen Qi Xiao, 1984, cited in Lindert, 1996a) show that since the 1930s, urban industrial expansion has resulted in an increase in the supply of manure as well as the supply of chemical fertilizers (through more favorable market prices), which has offset the decline in land availability. Lindert, Lee, and Wu (1996, cited in Lindert, 1996a) estimated that the net effect of nutrition losses from land loss and positive gains from urban and industrial expansion during 1930 to 1980 was positive.

II STATUS OF THE NATURAL RESOURCE BASE

The use of natural resources by humans, influenced by policy and governing institutions, can affect both the quality and quantity of these resources, which in turn will affect agricultural sustainability. Although the following investigation discusses each resource separately, the effects of agricultural activities on these resources often interact and overlap. For instance, salinization may result from mismanagement of water resources, but its impact is on land resources. Forests encompass a large number of resources including water, land, and living organisms. Deforestation increases the land available for agriculture, but reduces biodiversity and water quality, and may increase flooding and sedimentation.

LAND AND SOIL RESOURCES

The term "land" as used in this section refers to arable land or land cultivated with crops. It also includes land left fallow or used for pasture for less than five years (Engelman and LeRoy, 1995). The debate on land resources revolves around the technicalities of estimating the extent of their degradation.

Land Availability

The area of land per capita in Asia at present is lower than that in the Americas and Australia. Further, the proportion of soil highly suitable for cultivation in Asia is less than 4 percent, while in Latin America it is 12 percent and in Africa, 15 percent (Lohani, 1998).

Land suitability for agriculture depends on physical and chemical characteristics, but access and population determine whether land can be economically brought into agricultural production. Given that Asia has the highest population density of the five inhabited continents, pressure on land in this region is immense.

The area of arable land per capita reflects the population pressure on the land, and is an indicator of land stress (Table II.1). In 1992, arable land per head in Asia averaged 0.3 ha, which is considerably lower than the average of 1.6 ha for all other developing countries (Lohani, 1998). A critical threshold level is estimated to be 0.07 ha per capita (Smil, 1987, cited in Engelman and Leroy, 1995). This benchmark is derived from

Table II.1: Arable Land Scarcity Index (ha per capita) in Asia

	Year		
	1961	**1990**	**2025**
East Asia			
China, People's Rep. of	0.16	0.08	0.06
Japan	0.06	0.04	0.04
Korea, Rep. of	0.08	0.05	0.04
Southeast Asia			
Cambodia	0.43	0.35	0.16
Indonesia	0.18	0.12	0.08
Lao PDR	0.38	0.20	0.09
Malaysia	0.49	0.27	0.15
Myanmar	0.47	0.27	0.13
Philippines	0.24	0.13	0.08
Thailand	0.43	0.41	0.31
Viet Nam	0.17	0.10	0.05
South Asia			
Afghanistan	0.71	0.54	0.18
Bangladesh	0.17	0.09	0.05
India	0.36	0.20	0.12
Nepal	0.19	0.14	0.07
Pakistan	0.34	0.17	0.07
Sri Lanka	0.16	0.11	0.08

Source: Engelman and LeRoy (1995).

the area of arable land that would be able to feed a population sustainably, on a vegetarian diet basis, without the assistance of agrochemicals. A closed system with mixed cropping, crop recycling, and utilization of animal and human waste for maintaining soil fertility was assumed.

On the basis of a medium population projection estimated by the United Nations, Engelman and Leroy (1995) calculated the arable land scarcity index for 125 nations with populations of more than one million for 1960,1990, and 2025 (Table II.1). Japan dropped below the critical threshold in the early 1960s, followed by the Republic of Korea in 1990. By 2025, many Asian nations including PRC, Indonesia, Philippines, and Viet Nam will have dropped below the benchmark. The arable land scarcity index for Asia for 2025 is 0.12, substantially above the threshold level. By 2025, the Asian country with the most arable land per capita will be Thailand (0.31).

Land has always been a major constraint in agricultural production and was a major instigator of the green revolution. However, land availability is not the only factor that determines sustainability. The PRC has been able to feed its burgeoning population on a nonvegetarian basis despite the fact that land availability has fallen below the threshold level and that only 9.7 percent of the total land is arable (APO, 1998). In Viet Nam, where agricultural land per capita is amongst the lowest in Asia (0.10 ha in 1990), institutional reform has fueled a dramatic increase in output and enabled Viet Nam to become a major rice exporter.

Land Degradation

With proper care and management, land resources are renewable. However, under continuous degradation, the land could ultimately become nonproductive. Degradation may be the result of erosion, nutrient depletion, and/or physical and chemical contamination. Estimates of the quantity of land eroded each year range from 25 billion t (Pimentel et al., 1995) to 75 billion t (FAO, 1992, cited in Engelman and Leroy, 1995); most serious studies tend to confirm the lower estimates.

Experience worldwide, particularly in Africa, suggests that water and wind erosion accounts for the bulk of land degradation: 56 percent from water erosion, 28 percent from wind erosion, 12 percent from chemical degradation, and 4 percent from physical degradation (Oldeman, Hakkeling and Sombroek, 1991, cited in Crosson, 1994).

In nonirrigated areas, 85 percent of total erosion is brought about mainly by the effect of wind or water (Oldeman, 1992). On grazing land, degradation is often a result of overexploitation of public land, a process widely known as the "tragedy of the commons" (Hardin, 1968). In irrigated areas, chemical degradation results from mismanagement, high precipitation rate, soil type, topography, and population pressure. Global estimates by Dregne and Chou (1992, cited in Crosson, 1994) of land degradation suggest that its severity is greater in rangelands than in irrigated and rainfed croplands.

According to many estimates, agricultural land in Asia is extensively degraded, but the quoted data are not consistent. It was estimated in an FAO study (FAO, 1995a, p. 46) that of the 747 million ha of Asian cropland, 440 million ha (59 percent) have been degraded by water erosion, 222 million ha (30 percent) by wind erosion, 73 million ha (10 percent) by chemicals and 12 million ha (1 percent) by physical degradation. The "World Map of the Status of Human-Induced Soil Degradation", widely cited by the United Nations Environment Programme, the United Nations Development Programme, and the World Resources Institute, among others (Oldeman et al., 1990), indicates that 200 million ha of Asia's cropland and 200 million ha of rangeland have been degraded.

A more recent estimate of "human-induced" degradation (UNEP, 1997) shows 350 million ha as having been affected by topsoil loss, another 180 million ha by fertility decline, and 44 million ha by salinization. The most recent estimate (ADB, 1997a) indicates that 130 million ha of Asia's cropland have been salinized by poor irrigation practices, and 63 million ha of rainfed land and 16 million ha of irrigated land have been lost through desertification. This report has been summarized in a more recent paper (Crasswell, 1998): "the Asian Development Bank (ADB, 1997a) estimates that during the past 30 years one third of the agricultural land in Asia has been degraded".

A number of authors have questioned the accuracy of these estimates (Alexandratos, 1995; Crosson, 1995), which include the Global Land Assessment of Degradation. Crosson (1994) commented that most studies are based on "informed" opinions. Considering that enormous efforts have been made to rehabilitate watersheds, in India for example (Kerr et al., 1998), priority should be given to funding for scientific studies to enable a consensus on the extent of the degradation to be reached among international scientists.

At the national level, opinions on the extent of degradation also diverge and the examples quoted by opposing groups are not exactly comparable. Those who have raised concerns about the land degradation problem have pointed out that the amount of topsoil removed by runoff annually in India is around 25 billion t (Repetto, 1994, p.37). Nutrient depletion by loss of this topsoil is estimated to be equivalent to the total quantity of chemicals used over the entire country. Repetto (ibid.) further claimed that the loss in soil fertility has offset the yield improvement impact of the agricultural technology packages of India's 44 specialized research institutions in 26 States. Crosson (1994), citing a study by Bronger and Bruhn (1988, p. 688), argued that water erosion on the "red soils" that span over 700,000 km^2 (40 percent) of India's agricultural land, removes about 2.5 cm of topsoil per 100 years, which suggests that the impact under traditional agriculture was quite low. Both Repetto and Crosson did agree on one point: the cost of restoring land would be massive.

Outside India, erosion from the Loess Plateau in the PRC is often cited as an extreme example of soil erosion. Forty years of experience in soil conservation in this plateau have shown that effective control of sedimentation in the Yellow River requires integrated measures and that the process is costly and time consuming (Chen, 1992). In Southeast Asia, Indonesia, Lao PDR, Philippines, and Viet Nam tend to face more serious erosion problems because of their particular combinations of rainfall and topography.

While knowledge of the extent of degradation is useful for attracting the attention of policymakers worldwide, it would be more useful to know a) the relationship between degradation and yield loss, or the cost of degradation; b) the cost of stopping degradation; and c) the cost of rehabilitating the natural resources. Unfortunately, the answers to these questions are even more sketchy and uncertain. Studies are often based on very local conditions and are related to specific practices. Yield losses also vary from location to location. Moreover, the time frames over which these losses occur are not given, making it difficult to estimate the cost of erosion. If soils have become as degraded as the above estimates have intended to demonstrate, how can Asia's enormous food production, that feeds billions, be explained?

Among the few rigorous studies on the impact of soil erosion on yields is a study in the USA indicating that the effect of erosion-induced loss of soil productivity on corn and soybean yield there was very small and that the effect on wheat yields was negligible (Crosson, 1995). Soil erosion has caused yield losses of about 4 percent during the past 100 years. Based on the studies of both Dregne and Chou (1992) and Oldeman et al. (1990), Crosson (1995) concluded that the gain in food production from restoring land, or from attempting to reduce soil nutrient depletion, is negligible.

Studies concerning the effects of yield losses from erosion in Asia have been site-, crop- and practice-specific and cannot be generalized. Long-term studies conducted in the PRC from the 1930s to the 1980s, and in Indonesia from the 1940s to the 1990s, which combined soil surveys and input variables, revealed that over the period studied, soil organic matter and nitrogen had declined but total phosphorus and potassium had increased (Lindert, 1996a). Salinity and acidity had not shown worsening trends. In the PRC, the long-term effects of soil on yields showed that depletion of soil organic matter and total nitrogen in the Huang, Huai, Hai, and Chang Jiang plains could be reversed easily by using a quick-release fertilizer. The overall output depends on soil properties, and these had not shown depletion symptoms since the 1930s. In the southern PRC, acidity was even reduced between the 1930s and 1950s. In Indonesia, a significant drop in organic matter content was observed in various types of cultivation, including tree crops, dry-land, field crops, and fallow, although the level of both potassium and phosphorus increased. Acidity levels were reduced over time on dry land, although less reduction was found for rice.

It is important to note that much of the land defined as degraded actually contributes marginally to total crop production. For Asia, the picture becomes clearer when the problem of degradation is assessed separately for the more and the less productive croplands.

The more productive lands that have largely contributed to agricultural growth since 1960, i.e. irrigated land and rainfed areas with reliable rainfall and good soil, have yet to demonstrate effects of degradation. The regionwide yields (per ha) of rice, wheat, and maize have shown steady linear increases ($R^2 = 0.96$–0.98) for almost 40 years. It has been claimed that intensification of the rice cropping system has brought about environmental degradation, such as a reduction in soil quality and fertility. However, while soil analyses over the past 15 years in Karnal in Haryana, India, indicate a significant decline in soil nutrient levels (Mehta, 1990, cited in Chand and Haque, 1997), a major decline in yield levels has not yet been detected. There is indeed land degradation caused by intensive monocropping, but this can be solved by good soil and crop management, as discussed later.

Much of Asia's land degradation is in the less favorable areas, which contribute less to total production relative to their total area than do more favorable croplands (e.g. for rice, the uplands and unfavorable rainfed lowlands account for only 18 percent of the region's production and cover 42 percent of its area). Losses in production due to land degradation in these less favorable environments are not likely to be very large in terms of total output when compared with national and regional totals. Degraded lands are, however, generally in the most poverty-stricken areas in each country (e.g. the erosion-prone uplands/highlands in most countries from Nepal to Indonesia; the salt-affected areas of India, Pakistan, and Central Asia; the dry lands in the northeastern PRC and Central Asia, and the Loess Plateau in the PRC). In such areas, the losses in crop productivity due to land degradation would mean loss of a significant portion of income for the poorer, if not the poorest segments of the economy.

Success stories from India and Sri Lanka have been highlighted recently by the Technical Advisory Committee of the CGIAR (TAC, 1997), and have illustrated that sustainable growth in crop production on degraded land is possible. These cases have, however, also shown that very different sets of technological and institutional innovations are required.

In many instances, the net erosion loss could be much less than that observed onsite. Unless the eroded soil causes siltation in water resources, a loss from one production region could be a gain elsewhere. River deltas all over the world are manifestations of this gain. Moreover, the silt-laden water of the Yellow River, for example, could also be used to reclaim desert land (Fullen and Mitchell, 1994).

Diversification from food crops into higher-value crops could take more land out of food crops than land degradation. This phenomenon is spreading rapidly in the southwestern and southern parts of the PRC, the Chao Phraya Delta in Thailand, and the Red River Delta in Viet Nam. It is not a major threat because as the supply of basic food crops falls, prices will rise and land will again be brought back into basic food crop production. Price increases will harm the poor, and the issue then becomes how to guarantee nutrition and provide food security for the poor. Public policies designed for poverty alleviation are necessary.

FOREST RESOURCES

The role and value of forests have undergone more dramatic changes than have those of any other natural resources. Forests were previously valued mainly for timber and as land reserves for agriculture. Sustainability issues have significantly changed the concept of the value of forests. Today, they are valued not only as a source of land, timber, and other forest products, but also for their ecological and social functions, e.g. regulation of stream flows, soil and water conservation, microclimate regulation, carbon sequestration, tourism, recreation, and as a store of future wealth in the form of biodiversity. Forests are increasingly viewed as important sources of available and untapped genetic resources.

The multiplicity of forest resources and their uses has led to substantial conflicts both at the policy and grass-roots level. Forests are viewed by the State in terms of potential development projects and by farmers as potential farmland. NGOs, environmentalists, water users, and urban residents are demanding that forests be protected. While the debate related to land resource degradation is mainly scientific, issues related to forests are more complicated and have greater social ramifications because decisions concerning forests may have an impact on the livelihoods of the millions of persons residing in and around them.

Status of Asian Forests

About one quarter of the world's total land is presently forested (FAO, 1997a). In Asia, the proportion is only about 17 percent. The total forest area in Asia in 1995 was estimated at 499 million ha (FAO, 1997a). The Asian region also has the lowest ratio of forestland per capita (0.1 ha), much less than the world 's average of 0.6 ha. Mongolia, which is one of the least forested countries measured in terms of forest area to total land area, has the highest forest area per capita (4 ha) in Asia.

The proportion of forest in total land area is highest in the most and the least developed countries (Table II.2), e.g. Republic of Korea (77 percent), Japan (67 percent), Bhutan (59 percent), and Cambodia (58 percent). As a subregion, Southeast Asia contains the largest area of natural forests in Asia.

Changes in Forest Cover

The world's forested area decreased at the rate of 0.3 percent (or 11.3 million ha) per year during 1991–1995. There was a marked contrast in deforestation rates between developed and developing regions. In developed regions such as Europe and North America (Canada and USA), the forested areas increased at an annual rate of 0.1–0.3 percent (0.4–0.6 million ha) during 1991–1995. Conversely, the deforested areas of the developing tropical regions ranged between one and five million ha. The annual rate of forest decrease was 0.7 percent for tropical Asia (3,328,000 ha), 0.7 percent for tropical Africa (3,695,000 ha), 1.2 percent for Central America and Mexico (959,000 ha), and 0.6 percent for tropical South America (4,655,000 ha) (FAO, 1997a, Annex 3, Table 3, p. 186-189).

In Asia, the annual rate of deforestation is highest in Southeast Asia, at about 1.3 percent or 2.9 million ha, and lowest in East Asia, 0.1–0.2 percent per year. In contrast, the forested areas in Central Asia increased markedly during this period (Table II.3). The countries where the annual rate of deforestation in 1991–1995 was faster than that in 1981–1990, or with annual deforestation rates greater than 2 percent are Afghanistan, Cambodia, Kazakhstan, Malaysia, Pakistan, Philippines, Thailand, and Uzbekistan. It should be noted that most of these countries do not have high population pressure.

The causes of deforestation are complex and numerous. A typical pattern in the humid tropics generally starts with unsustainable logging, which helps clear the forest for slash-and-burn agriculture or provides easier access for commercial agriculture. Weak administration and corruption have rendered demarcation, monitoring, and enforcement ineffective, and

Table II.2: Forest Resources in Selected Asian Economies, 1995

	Land Area (ha'000)	Population (million)	Forest Area 1995 (ha'000)				
			Total	(% land)	ha/cap.	Natural Forest	Plantation
East Asia							
China, People's Rep. of	932,641	1,221.5	133,323	14.3	0.1	99,523	33,800
Japan	37,652	125.2	25,146	66.8	0.2	25,146	nc
Korea, Rep. of	9,873	45.0	7,626	77.2	0.2	6,226	1,400
Mongolia	156,650	2.4	9,406	6.0	3.9	9,406	0
Southeast Asia							
Cambodia	17,652	10.3	9,830	55.7	1.0	9,823	7
Indonesia	181,157	197.6	109,791	60.6	0.6	103,666	6,125
Lao PDR	23,080	4.9	12,435	53.9	2.5	12,431	4
Malaysia	32,855	20.1	15,471	47.1	0.8	15,371	100
Myanmar	65,755	46.5	27,151	41.3	0.6	26,875	276
Philippines	29,817	67.6	6,766	22.7	0.1	6,563	203
Thailand	51,089	58.8	11,630	22.8	0.2	11,101	529
Viet Nam	32,549	74.5	9,117	28.0	0.1	7,647	1,470
South Asia							
Afghanistan	65,209	20.1	1,398	2.1	0.1	1,390	8
Bangladesh	13,017	120.4	1,010	7.8	nc	700	310
Bhutan	4,700	1.6	2,756	58.6	1.7	2,748	8
India	297,319	935.7	65,005	21.9	0.1	50,385	14,620
Maldives	30	0.3					
Nepal	13,680	21.9	4,822	35.2	0.2	4,766	56
Pakistan	77,088	140.5	1,748	2.3		1,580	168
Sri Lanka	6,463	18.4	1,796	27.8	0.1	1,657	139
Central Asia							
Kazakhstan	267,073	16.5	10,504	3.9	0.6	nc	nc
Kyrgyz Republic	19,180	4.5	730	3.8	0.2	nc	nc
Tajikistan	14,060	5.8	410	2.9	0.1	nc	nc
Turkmenistan	46,993	4.5	3,754	8.0	0.9	nc	nc
Uzbekistan	41,424	22.8	9,119	22.0	0.4	nc	nc

nc = data not classified.

Source: Modified from FAO (1997a).

Table II.3: Changes in Forest Cover in Selected Asian Economies, 1980–1995

	Forest Area (ha'000)			Annual Change (ha'000)		Annual Change (%)	
	1980	1990	1995	1981–1990	1991–1995	1981–1990	1991–1995
East Asia							
China, People's Rep. of	137,756	133,756	133,323	-400	-87	-0.3	-0.1
Japan	25,262	25,212	25,146	-5	-13	0.0	-0.1
Korea, Rep. of	7,701	7,691	7,626	-1	-13	0.0	-0.2
Mongolia	9,406	9,406	9,406	0	0	0.0	0.0
Southeast Asia							
Cambodia	11,959	10,649	9,830	-131	-164	-1.1	-1.5
Indonesia	127,333	115,213	109,791	-1,212	-1,084	-1.0	-0.9
Lao DPR	14,467	13,177	12,435	-129	-148	-0.9	-1.1
Malaysia	21,432	17,472	15,471	-396	-400	-1.8	-2.3
Myanmar	33,098	29,088	27,151	-401	-387	-1.2	-1.3
Philippines	11,238	8,078	6,766	-316	-262	-2.8	-3.2
Thailand	18,427	13,277	11,630	-515	-329	-2.8	-2.5
Viet Nam	11,163	9,793	9,117	-137	-135	-1.2	-1.4
South Asia							
Afghanistan	1,990	1,990	1,398	0	-118	0.0	-5.9
Bangladesh	1,434	1,054	1,010	-38	-9	-2.6	-0.9
Bhutan	2,963	2,803	2,756	-16	-9	-0.5	-0.3
India	68,359	64,969	65,005	-339	7	-0.5	0.0
Nepal	5,636	5,096	4,822	-54	-55	-1.0	-1.1
Pakistan	2,793	2,023	1,748	-77	-55	-2.8	-2.7
Sri Lanka	2,167	1,897	1,796	-27	-20	-1.2	-1.1
Central Asia							
Kazakhstan		9,540	10,504		193		2.0
Kyrgyz Rep.		730	730		0		0.0
Tajikistan		410	410		0		0.0
Turkmenistan		3,754	3,754		0		0.0
Uzbekistan		7,989	9,119		226		2.8

Source: modified from FAO (1995b, 1997a).

public forestland is turned by the powerful into private domains (Box II.1). The existence and contribution of farmers' settlements in the forests are often ignored, and recently in some countries serious and violent conflicts between forest communities and the State have become frequent. Improved infrastructure and health services, reduced death rates, and improved living conditions encourage an increase in population through both natural growth and migration. In areas where the scenery is favorable, economic booms may induce conversion of forestland into tourist resorts or large-scale plantations.

Deforestation accelerates soil erosion because runoff increases, leading to increased sedimentation. Unsustainable logging and the conversion of forests into cropland, e.g. coffee, rubber, and banana in the northern Mekong Delta and Nambo region of Viet Nam, have been major causes of erosion and consequentially of sedimentation. Forest cover decreased from 70 percent of the total area in the 1940s to less than 30 percent in the 1990s, an annual reduction rate of 1.6 percent (Crooks, 1995).

Excessive siltation in reservoirs is reported in India, where actual siltation rates in 12 major dams exceeded the designed siltation capacity by a factor of two, and thus considerably shortened the useful life of the dams (Repetto, 1994). In Viet Nam, the most cited case is the Hua Binh Dam; heavy sedimentation threatened to reduce its useful life. The problem led the Government to build an upstream dam to reduce the sediment load (Mie Xie, 1996).

Biodiversity

Forest resources include plant and animal ecosystems as well as physical assets such as land and water. Plant and animal biodiversity is generally greatest in tropical forests.

Diversity is a fundamental characteristic of sustainability. The wild-growing relatives of crop species protect crop varieties against extinction from pest or disease outbreak. Apart from their use in direct consumption, wild plants and animals are

Box II.1 Poverty, Power, and Public Resources

A study of forest use in Thailand during the economic boom revealed an interesting pattern of forest usage by the elite (Anan and Mingsarn, 1996). Starting in 1981, the Thai economy grew at double-digit rates consecutively for three years, creating high expectations for the country's future. In a country where loans are made on the basis of collateral rather than financial feasibility of a project, the demand for land for collateral increased rapidly, driving the price of titled land to great heights.

The highlands, which are mostly fragile ecosystems, are much coveted as they can be used as sites for tourist resorts or for tourism-cum-agriculture projects. In addition, returns from subtemperate agriculture and highland agriculture are artificially high owing to government protection from imported agricultural produce. The lucrative returns from both lowlands and highlands have encouraged encroachment on forests in fragile ecosystems by the urban elite.

The urban elite would not have been able to locate suitable or scenic upland and highland sites on their own. The purchase of this forestland has been possible due to the flow of information between local leaders, who are vote collectors for national politicians, and their supporters, who are wealthy businesspersons. Another important condition that has made encroachments and purchases worthwhile is the ability to convert public land into private land through the abuse of administrative power, a situation that encouraged national and local politicians to join hands in converting forestland. Poor enforcement of forest laws also allowed rich people with less influence to pay for the de facto usufruct rights given to the early encroachers.

A previous Thai Government allocated forestland on the assumption that the de facto owners of these public lands were the landless. That Government was toppled when it was found that some of the land grantees were the richest people in the province. Policies based on the assumption that only the poor use forests are destined to fail. The various forest resources are

(continued next page)

(Box II.1 continued)

extracted by people of varying income and power. Different policy instruments are needed in order to deal with these different groups. For the powerful elite, increased public awareness and "people power" have proven, at least in Thailand, able to counter the abuse of power and to topple a government known to be involved in forestland scandals. Tax instruments are needed to deal with wealthy land seekers while for the poor, who encroach on forests out of necessity, the need is for policy instruments that protect the environment and include poverty alleviation as a joint objective. Similarly, policies aimed at poverty reduction need to recognize environmental constraints in order to achieve sustainable development.

important sources of scientific information. Genetic information from some of these resources may reveal properties that increase the immunity of crop varieties to certain diseases. For example, genes from a wild rice species from India provide immunity in cultivated rice to four types of disease (Bryant, 1998). Desirable properties can be synthesized and need not be directly extracted from the materials.

Currently, knowledge of potential uses of the genetic information stored in forest resources is negligible. Considering flowering plants alone, less than 10 percent of the known 250,000 species have been scientifically studied. With forests the size of Cambodia disappearing every year (Reid, 1993), the world's options for a better future are definitely at stake.

Six Asian countries, PRC, India, Indonesia, Malaysia, Philippines, and Viet Nam are among the 20 "mega-diversity" centers of the world (Paine, Byron, and Poffenberger, 1997). Animal and plant species in these six countries account for almost 60 percent of total world species.

Despite its relatively small proportion of forestland, the PRC is ranked as eighth in the world in terms of biodiversity, and first in the northern hemisphere. The country is endowed with 32,800 plant species and 104,500 animal species. Some 200 species of plants have disappeared, while a further 5,000 are endangered (Yin Runsheng, 1997). Indonesia, which has only 1.7 percent of the world's land area, accounts for 17 percent of all plant and animal species or more than the known species of the whole of Africa (State Ministry of Environment and KONPHALINDO, 1995).

The threat to biodiversity is often not from agriculture, but is related to the lack of or inappropriate management. A recent study has shown that in some countries where biodiversity is relatively great, such as India, Indonesia, Malaysia, Myanmar, and Viet Nam, the level of protection is not very strong. Often this is simply the result of the perception that where biodiversity is strong there is less need for protection (Paine, Byron, and Poffenberger, 1997).

Although wild genetic resources are potentially important to maintain varietal diversity, conservation is often costly even in the natural habitat. For major food crops, an international system exists within the CGIAR for conserving plant genetic materials. It is argued that the germplasm of major food crops is reasonably protected under this system although greater efforts could be made to collect wild species (Hawkes, 1985; Crosson and Anderson, 1985, cited in Crosson, 1994). While the genetic variation in landraces is well protected in seed banks, most of the accessions provide little information and have not been not evaluated in germination tests (McNeely et al., 1990). Crosson (1992) concluded that since the viability of the seed bank depends on the viability of the CGIAR system, which is constantly financially insecure, the threat to compiling and maintaining knowledge on genetic resources is probably bigger than the threat to the natural resource base itself.

There have been several attempts to value biodiversity loss in Asia. In the 42-km^2 Yom basin, Thailand, the loss in genetic value of the teak forests that are to be removed to make way for a reservoir is estimated at $60 million. The realizable annual ecological benefits of the mangroves of Bantuni Bay, Indonesia, are valued at $1,500 per km^2, while those of tropical forests are estimated at $3,000 per km^2 (Ruitenbeek, 1990, cited in Bann, 1998).

It should be noted that the biodiversity loss that will affect agriculture, especially food security, is more related to diversity within species of major staples than to biodiversity in general, although genetic engineering will make cross-species gene transfers more feasible (Evenson, 1996). Therefore, options for managing international gene pools are mostly limited to gene banks and habitat protection; there are a greater number of options for financing habitat protection.

Protected Areas

Many countries attempt to protect biodiversity using protected area regimes. A protected area is "an area of land or sea especially dedicated to the protection and maintenance of biological diversity, and of natural and associated cultural resources, and managed through legal or other effective means" (McNeely, Harrison, and Dingwall, 1994). There are six categories of protected areas, related to the different objectives of their establishment. Since 1970, State acquisition of protected areas has increased markedly. Countries with the highest proportion of protected land areas are Bhutan (21 percent), Cambodia (16 percent), Thailand (13 percent), and Indonesia (10 percent) (Table II.4). Except for Indonesia, the Asian countries designated as mega-centers of biodiversity tend to have relatively low ratios of protected area to total land area.

An important feature of protected areas in Asia is that they are not areas of pure wilderness, and they are occasionally inhabited by indigenous communities that have traditionally relied on forest resources for livelihood. Often the modern legal instruments used by the State to claim protected areas have neglected the rights of traditional users, depriving local communities of their usual sources of sustenance, and resulting in bitter and at times violent conflicts between the State and local communities.

Table II.4:   National and International Protected Areas in Asia, 1997

	Land Area (ha'000)	National Protected Areas IUCN Class I–V			International Protected Areas					
					Biosphere Res.		World Heritage		Wetland	
	(ha'000)	No.	(ha'000)	%	No.	(ha'000)	No.	(ha'000)	No.	(ha'000)
Asia	2,442,538	1,490	143,367	5.9	36	10,994	21	1,853	44	2,738
East Asia	1,142,245	409	79,483	7.0	20	8,166	8	252	18	672
China, People's Rep. of	932,641	265	59,807	6.4	12	2,514	6	224	7	588
Japan	37,652	65	2,550	6.8	4	116	2	28	10	84
Korea, Dem. People's Rep. of	5,400	19	315	2.6	1	132	0	0	0	
Korea, Rep. of	9,902	25	682	6.9	0		0	0	0	
Mongolia	156,650	35	16,129	10.3	2	5,367	0	0	0	
Southeast Asia	433,996	423	31,163	7.2	11	2,682	5	1,103	5	299
Cambodia	17,652	20	2,863	16.2	0		0	0	0	
Indonesia	181,157	170	17,509	9.7	6	1,482	2	298	2	243
Lao PDR	23,080				0		0	0	0	
Malaysia	32,855	50	1,483	4.5	0		0	0	1	38
Myanmar	65,797	2	173	0.3	0		0	0	0	
Philippines	29,817	17	1,453	4.9	2	1,174	1	33	1	6
Singapore	61	1	3	4.4	0		0	0	0	
Thailand	51,089	112	6,688	13.1	3	26	1	622	0	
Viet Nam	32,549	52	994	3.1	0		1	150	1	12
South Asia	477,506	504	21,279	4.5	3	40	8	498	17	339
Afghanistan	65,209	6	218	0.3	0		0	0	0	
Bangladesh	13,017	9	98	0.8	0		0	0	1	60
Bhutan	4,700	9	998	21.2	0		0	0	0	
India	297,319	344	14,273	4.8	0		5	281	6	193
Maldives	30									
Nepal	13,680	12	1,112	7.8	0		2	208	1	18
Pakistan	77,088	55	3,721	4.8	1	31	0	0	8	62
Sri Lanka	6,463	69	859	13.3	2	9	1	9	1	6
Central Asia	388,730	153	11,439	2.9	2	106	0	0	4	1,428
Kazakhstan	267,073	70	7,337	2.7	0		0	0	2	609
Kyrgyz Republic	19,180	31	688	3.6	1	71	0	0	1	630
Uzbekistan	14,060	12	850	2.1	0		0	0	0	
Tajikistan	46,993	18	587	4.2	0		0	0	0	
Turkmenistan	41,424	22	1,977	4.2	1	35	0	0	1	189

Source: WRI (1997).

Note: The protected area management systems are as follows.

IUCN has six categories (McNeely, Harrison, and Dingwall, 1994): I Strict Nature Reserve/Wilderness Areas; II National Parks–protected areas managed mainly for ecosystem conservation and recreation; III Nature Monuments–protected areas managed mainly for conservation of specific features; IV Habitat/Species Management Areas–protected areas managed mainly for conservation through management intervention; V Protected Landscapes/Seascapes–protected areas managed mainly for landscape/seascape conservation and recreation; VI (not used in Table) Managed Resources Protected Areas–protected areas managed mainly for sustainable use of natural ecosystems.

National protection systems: all protected areas including all those as classified by IUCN (I-V), as totally protected areas IUCN (I-III) and partially protected areas IUCN (IV-V).

Biosphere reserves: representative terrestrial and coastal environments that have been internationally recognized under UNESCO's Man and Biosphere Programme.

World Heritage sites: areas of outstanding universal value, or with both natural and cultural values.

Wetlands of international importance: areas declared and recognized by the Convention on Wetlands of International Importance (Ramsar, Iran, 1971). There are 44 sites in Asia.

Laws covering protected areas are more stringent than forestry laws; consequently protected areas have become legal mechanisms used by the State for protecting public lands (as opposed to biodiversity) from encroachment. Most protected areas are established on an ad hoc rather than a scientific basis. The overexpansion of areas under protection renders that protection infeasible, and when this is coupled with poor human resource development in conservation, the result is often a "paper park", an area protected by law in theory but not in practice.

During the 1990s, State acquisition of protected areas has slowed down. This is partly because for some countries, there are no more areas suitable for protected status. Social conflicts on land rights with local communities have also slowed the process of designation. The cost of protection is also relatively high in Southeast Asia where natural forests are dwindling more rapidly. The annual cost of protection per km^2 in Southeast Asia is \$509, compared with \$175 and \$359 in South Asia and East Asia, respectively (Paine, Byron, and Poffenberger, 1997). The average staff to area ratio also varies considerably from 26 and 29 per 1,000 km^2 in East Asia and Southeast Asia, respectively, to 81 in South Asia; the worldwide average is 25 staff per 1,000 km^2.

Most protected areas contain scenic beauty and ample opportunities exist to finance protection through tourism, for example by charging appropriate entrance fees. There is a broad range of alternative financing options for protected areas, including debt-for-nature swaps, bioprospecting, conservation funds, and nondevelopment rights (Reid et al.,1993; Mingsarn et al., 1995).

WATER RESOURCES

Unlike land, where the debate focuses on technical estimates of the size of the problem, there is general agreement that water will be increasingly diverted from agriculture to other, high-value uses. Controversies related to water center around how this can be done in an efficient and equitable manner, for example by minimizing the impact of large-scale water resource development projects on local communities and the environment, and the appropriate pricing of water without undue political consequences.

Although water is a renewable resource, the maximum amount that can be used during each season is fixed according to the geography and climate of each location. The water situation in a given area can be roughly assessed by a water availability indicator: the amount of annually renewable water per capita. The renewable water supply includes both internally produced surface water and groundwater and river flow from external sources. A threshold level for water scarcity of 1,000 persons per one flow unit (1 million cubic meters (m^3) per year), equivalent to 1,000 m^3 per capita per year, has been proposed (Falkenmark, Lundqvist, and Widstrand, 1989). A country can be described as facing a water stress situation if the water supply is between 1,000 and 1,700 m^3 per capita (Fig. II.1). Below a

Water Stress Index

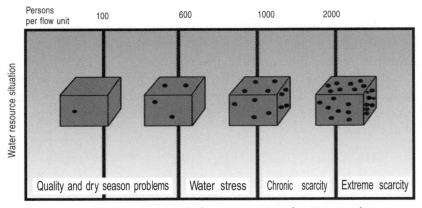

Figure II.1: The water stress index, a measure of water scarcity

Note: 1 flow unit = 1 million cubic meters per year

Source: Adapted from Falkenmark, M. 1991 The Ven Te Chow Memorial Lecture. *Water International* 16 (4): 229-240.

level of 1,000 m^3 per capita, the population faces extreme water scarcity (Table II.5). The estimated annual minimum water requirement for basic needs ranges from 400 to 2,000 m^3 per capita (Rosegrant and Ringler, 1998).

Population increase is a major cause of water stress. The per capita availability of water in Asia is the lowest among all

Table II.5: Water Resources in Asia

				Annual Withdrawals			
	Annual Internal Renewable Water Resources[a]		Annual River Flows From	Volume	Proportion of Water Resource		Per Capita
	Total (km^3)	1998 per capita (m^3)	External Sources (km^3)	(km^3)	Internal sources only (%)	Including external sources (%)	(m^3)
East Asia							
China, People's Rep. of	2,800.00	2,231.00	0.00	460.00	16.43	16.43	461.00
Japan	547.00	4,344.00	0.00	90.80	16.60	16.60	735.00
Korea, Rep. of	66.12	1,434.00		27.60	41.74	41.74	632.00
Mongolia	24.60	9,375.00		0.55	2.25	2.24	271.00
Southeast Asia							
Cambodia	88.10	8,195.00	410.00	0.52	0.59	0.10	66.00
Indonesia	2,530.00	12,251.00	0.00	16.59	0.66	0.66	96.00
Lao PDR	270.00	50,392.00		0.99	0.37	0.37	259.00
Malaysia	456.00	21,259.00		9.42	2.07	2.07	768.00
Myanmar	1,082.00	22,719.00		3.96	0.37	0.37	101.00
Philippines	323.00	4,476.00	0.00	29.50	9.13	9.13	686.00
Singapore	0.60	172.00	0.00	0.19	31.67	31.67	84.00
Thailand	110.00	1,845.00	69.00	31.90	29	17.82	602.00
Viet Nam	376.00	4,827.00		28.90	7.69	7.69	416.00
South Asia							
Afghanistan	55.00	2,354.00	10.00	25.85	47.00	39.77	1,825.00
Bangladesh	1,357.00	10,940.00	1,000.00	22.50	1.66	0.95	217.00
Bhutan	95.00	49,557.00		0.02	0.02	0.02	13.00
India	1,850.00	1,896.00	235.00	380.00	20.54	18.23	612.00
Nepal	170.00	7,338.00		2.68	1.58	1.58	154.00
Pakistan	248.00	1,678.00	170.30	155.60	62.74	37.20	1,269.00
Sri Lanka	43.20	2,341.00	0.00	6.30	14.58	14.58	503.00
Central Asia							
Kazakhstan	75.42	4,484.00	34.20	33.67	44.64	30.72	2,002.00
Uzbekistan	16.34	704.00	34.10	58.05	355.26	115.09	2,501.00
Tajikistan	66.30	11,171.00	50.30	11.87	17.90	10.18	2,001.00
Turkmenistan	1.00	232.00	70.00	23.78	2,378.00	33.49	5,723.00

0 = zero or less than half of the unit measured.
[a] Annual internal renewable water excludes river flows from other countries.
[b] Water withdrawal includes all water used for irrigation, industry, and agriculture (including watering of animals). It is not the same as water consumption.

Source: WRI (1998).

the continents (ADB, 1998); the water availability indicator dropped from 9,600 m^3 in 1950 to 3,240 m^3 in 1992 (WRI, 1994, Table 22). In rapidly growing countries, the competition for water between agriculture and other uses is likely to increase. Projections of water demand to the year 2025 reveal a switch in the demand pattern to more nonfarm uses. The growth in nonfarm use is particularly strong in PRC, India, and Southeast Asia (Rosegrant and Ringler, 1997).

On the basis of this indicator, Singapore, Turkmenistan, and Uzbekistan face extreme water scarcity conditions. However, the indicator roughly indicates overall supply only. Care must be taken when it is used to compare countries with different climatic conditions. In tropical countries, evaporative loss is much higher than in temperate countries, and the same level of precipitation may imply a very different water situation. In addition, the distribution of rainfall over the year and the length of the rainy season also make a difference in the water situation given the same level of availability. Further, the extent and severity of water pollution need to be considered.

The above supply-side indicator is more useful for guiding water management strategies when used in combination with a water withdrawal indicator (sometimes referred to as the United Nations indicator) which indicates the proportion of available water used (Table II.5). Water management problems are relatively easy to deal with if the withdrawal indicator is below 10 percent of availability (Falkenmark, Lundqvist, and Widstrand, 1989). Above 20 percent, water management becomes a major concern. At relatively high levels of water utilization, as the gap between availability and use further declines, water management strategies have to include storage enlargement, rationing, and conservation. In this case, water has to be withdrawn from outside sources. Countries with utilization rates above 40 percent are considered to be experiencing high levels of water stress. Under this indicator (Table II.5), Afghanistan, Central Asia, and the Republic of Korea appear to have major water problems. In Central Asia, the withdrawal ratio is very high in Kazakhstan, Uzbekistan, and Turkmenistan. This is mainly the result of faulty irrigation systems with canals that lose most of their water on the way to target areas. If external water sources are excluded, Thailand and India would be in relatively vulnerable water positions.

In Asia, Southeast Asia tends to be most abundantly endowed with water and the level of utilization is low, except for Singapore (Table II.5). The major problems in the deltas are drainage and eutrophication, due to the relatively humid climate. In fact, in the Mekong and Irrawaddy deltas, efficient water management and drainage systems have higher priority than large-scale water infrastructure projects.

The water availability and withdrawal indicators can be misleading when water is very unevenly distributed within a country. For example, in the PRC, water is most plentiful in the southeast where arable land is scarce. Water availability is low in the northern region above the Yangtze River basin where agriculture is intensive. While much of the surface water (80 percent) is from the Yangtze and other basins in the south, most of the cropland (63 percent) is north of the Yangtze (Brown, 1995). In addition, water can be unevenly distributed between seasons. Bangladesh and Viet Nam encounter both long periods of flooding and dry season shortages.

Also, the water availability and withdrawal indicators do not take future uses into account. The International Water Management Institute (IWMI) (Seckler et al., 1998) has attempted to provide an indicator that reflects water requirements in 2025, based on data in 118 countries. Two scenarios, business as usual and with more effective irrigation, are given. Countries were ranked into five groups according to two criteria: the percentage increase in water withdrawal between 1990 and 2025, and the extent of withdrawals in the year 2025. Group 1 consists of countries that are water scarce according to both criteria. Countries in group 2 need to develop water resources fully in order to meet increased requirements in 2025. The third group comprises countries that would need to develop 25–100 percent, 48 percent on average, of their water. Groups 4 and 5 consist of countries that have adequate water supplies, and for group 5, future requirements tend to decrease.

The IWMI model does not deal with changes in the crop and industrial mix.

Under the business-as-usual scenario, the projected increases in water withdrawals in Asian countries range from 40 to 135 percent. Under the second scenario, which assumes substantial improvements in irrigation, only two Asian countries, Singapore and Pakistan, were included in group 1. None were included in group 2. Cambodia, Indonesia, Malaysia, Myanmar, and Nepal were included in group 3, and the Philippines and Viet Nam were in group 4. Surprisingly, the Republic of Korea and Thailand, both of which are under water stress according to the United Nations indicator, and are emerging as countries facing severe or increasing water resource management problems, were ranked in group 5. The IWMI study treated the PRC and India separately, but if the same criteria were applied, the PRC would have been placed in group 5 and India in group 4.

At first glance, the IWMI indicator seems to provide a picture that conflicts with conventional indicators. However, this indicator is a projection to a future situation (year 2025), while the two indicators mentioned earlier reflect the existing situation. The IWMI projections assume increased irrigation effectiveness from around 43 percent on average currently to 60 percent in 2025. In the case of Thailand, the IWMI prediction tends to overstate the water situation because it uses the unrevised World Resources Institute data from 1990 on annual internal renewable resources, which are 69 percent greater than the revised figures for 1998. Also, the Thai projection is very sensitive to the assumption of increased irrigation efficiency because it has a very high proportion of water use for irrigation relative to other economic sectors than is the case in most other countries. Moreover, the IWMI indicator cannot properly reflect the water situation in the PRC where water is abundant where it is less needed. However, the IWMI prediction is a good illustration of the gains to be derived from improving irrigation effectiveness. According to IWMI, if Thailand can improve its irrigation efficiency by 100 percent, i.e., from 31 to 60 percent between 1990 and 2025, a substantial volume of water will be liberated from agricultural use. Total withdrawals in 2025 would then be lower than in 1990 by 11 percent without the need for new, major water resource development projects.

Pollution may make water unfit for higher-value uses. For example, pollution is emerging as a constraint to the expansion of Chinese inland fisheries. Another example is shrimp aquaculture, where shifting cultivation practices are dictated by the need for water of suitable quality in order to avoid overstocking and disease contagion.

Underground water is an important and reliable source of water for irrigation in many Asian countries. Bangladesh, Pakistan, and India all rely on groundwater for more than 30 percent of their total water resources. Recently, these resources have been placed under threat from overpumping. In India, where half the irrigation area is under pumping irrigation, groundwater is now a very valuable resource for agriculture. It is estimated that groundwater is used in 75–80 percent of Indian irrigation. Overexploitation of groundwater for rice and wheat cultivation has caused the water table to drop by 30–40 cm annually in the Gangetic Plain, and especially in Ludhaina, Paltana, and Sangur in Punjab, and Karnal in Haryana (Chand and Haque, 1997).

Overpumping across millions of hectares of the coastal areas of Gujarat and Tamil Nadu has caused seawater seepage into the aquifer (Repetto, 1994). A similar phenomenon is observed in the southern coastal areas of Viet Nam, leading to the abandonment of groundwater extraction works in the area. Open-access regimes have encouraged excessive water pumping and have lowered the water table substantially in Gujarat, Rajasthan, Punjab, and Karnataka (Moench, 1994). The water table dropped by 30 m in a few decades in Tamil Nadu (Falkenmark and Widstrand, 1992).

As stated earlier, water availability is considered to be an increasing constraint in the PRC. Prior to 1970, irrigation water there came from the development of surface sources. Since 1970, groundwater has been increasingly tapped, resulting in falling water tables, especially in Hebei Province, where the groundwater table dropped 70 m in the Cangzhou area over a period of 10 years (Zhang and Zhang, 1995). In the northern region of the PRC, where reliance is placed on groundwater because the distribution of surface water is uneven, more than 70 percent of the extractable groundwater has been utilized (Zhang and Zhang, 1995).

To ensure the sustainable development of groundwater resources, knowledge of the characteristics of the aquifers and their natural recharge rates is important. It is particularly important for Bangladesh, India, and Pakistan where private-sector investments in tube-well irrigation have proliferated, and in Thailand where the conjunctive use of surface water and groundwater is gaining in importance.

Like other natural resources, water has multiple uses and therefore many stakeholders. In certain instances, the situation is further complicated when water resources are shared by a number of countries. The various uses can be potential sources for conflict because water extracted for one consumptive use may deprive other uses, e.g. water absorbed by plants will not be available for household uses such as drinking or washing. Differences in the timing of water use may also generate conflicts between consumptive and nonconsumptive uses, e.g. hydropower and navigation. Throughout Asia, conflicts about water are growing as the demand for water increases in every economic sector.

In the past, such conflicts arose mainly from competition for water for alternative economic uses, such as between upstream and downstream users, between economic sectors, and between rural and urban uses. Traditionally, water projects have often been designed to divert water resources from their natural state for uses that bring increased economic benefits. Recently, concern over the diversion of water from its natural environmental uses to economic uses has been heightened. Environmental groups in Asia, which are growing in number, knowledge, and experience, are demanding greater public participation and more environmentally friendly approaches in the planning, decision making, implementation, and management of water resource development projects.

In the next decade, water resource management will become a major challenge for large countries with big water deficits, such as the PRC and India, as well as Thailand where rice is a major foreign-exchange earner. By 2010, water availability per capita per year from the major tributaries (except the Ping and Nan rivers) in the Chao Phraya Basin, which is the rice bowl of Thailand, will be below 1,500 m^3. Therefore, improvements to water institutions and the efficiency of the irrigation system are imperative.

AQUATIC RESOURCE SYSTEMS AND FISHERIES RESOURCES

Fisheries and aquatic resources and the consequences of their degradation have received little publicity, unlike biodiversity and forest resources. The lack of public awareness of the status of aquatic resources stems from the fact that their degradation is concealed under water and is not evident, either visually or via satellite monitoring. Moreover, because research and data collection in this sector are difficult and costly, the growth of scientific knowledge of aquatic resources has lagged behind that for the food crop, livestock, and forest sectors (ICLARM, 1999).

ICLARM has divided aquatic resource systems into 1) ponds; 2) reservoirs and lakes; 3) streams, rivers, and floodplains; 4) coastal waters including estuaries and lagoons; 5) coral reefs; 6) soft bottom continental shelves (i.e. shelves up to 200 m in depth); 7) upwelling shelves; and 8) open oceans. The total economic value of the ecological services rendered by these aquatic resources systems is estimated at $21 trillion (ICLARM, 1999). In Asia, areas with upwelling, a process through which nutrients from lower layers of the sea are brought to the surface, are small and occur mainly around the northwestern Indian Ocean and parts of Indonesia. Thus, upwelling shelves, and also open oceans, will not be emphasized here as they are only remotely related to rural Asia.

Asia has about 29 million ha of natural lakes and 5.5 million ha of reservoirs (ICLARM, 1999). There are also vast areas of wetlands. For example, 18 percent of the Ganges and 9 percent of the Mekong basins are wetlands. In Cambodia, Tonle Sap, the largest inland lake in the lower Mekong sub-basin, and its flood plains and adjoining river systems, can be considered the subregional hub for aquatic diversity. Although fishery stock assessments for this very important area are grossly inadequate, at least 215 species have been recorded in the catch (Royal Government of Cambodia, 1998). The productivity of Tonle Sap fisheries is estimated at 65 kg/ha/year, more than five times that of most other tropical freshwater bodies, which average around 12kg/ha/year. The annual catch from Tonle Sap has always exceeded that from Cambodia's marine fisheries, accounting for 50 to 70 percent of the total national catch. For example, the lake's annual catches from 1993 to 1995 ranged between 60,000 and 72,000 t, compared with around 30,000 t from marine fisheries over the same period.

The lower Mekong sub-basin is the habitat of at least 1,200 different species of fish, of which 400 species are economically important to local communities (MRC, 1997). The annual catch for the four countries it covers is estimated at 815,000 to 940,000 t (van Zalinge, 1998). Catches of two large migratory fish species, the giant Mekong catfish *(Pangasianodon gigas)*, and the giant Mekong barb *(Catlocarpio siamensis)*, have declined drastically. Similar trends are forecast for medium-sized fish. The Irrawaddy dolphin is being threatened by a tourism project near the Li Pi Falls. Multipurpose dams built in Thailand, Lao PDR, and Cambodia have blocked fish migration and disturbed spawning areas.

Coastal waters, including estuaries, lagoons, and mangroves, have higher productivity than offshore or freshwater systems (ICLARM, 1999). The ecological services provided by estuaries (mainly in nutrient recycling and food production) are estimated to be worth as much as $22,000 per ha per year (Constanza et al., 1977, cited in ICLARM, 1999). Mangroves provide shoreline protection from storms, winds, and waves; serve as nutrient filters, sediment sinks, energy sources, and habitats for a large number of species of marine and terrestrial flora and fauna; and are important to the livelihoods of small-scale fishers. In sheltered tropical coasts, the high productivity of mangroves is believed to contribute to marine ecosystems via nutrient transportation as well as through their nursery and habitat functions. It is estimated that the global value of ecosystem services provided by mangroves totals $9,900 per ha per year, and more than four fifths of this value comes from their services of waste treatment and disturbance regulation (ibid.).

In recent decades, mangroves have been rapidly destroyed to provide land for human settlements, ports and other infrastructure development, charcoal, and space for aquaculture. In the Philippines, a substantial proportion of the mangrove area has been converted into fishponds to raise milkfish. In Thailand, intensive shrimp farming spread rapidly during the 1980s and used about 30 percent of the country's mangrove areas, although they are not suitable for shrimp farming. However, due to the high returns in shrimp aquaculture, profit can be made after only a few years of operation, and it was relatively easy to set up farms in mangroves owing to a lack of property rights and weak enforcement of forest laws.

Coastal and inland aquatic resources are now seriously threatened by land-based pollution. In the PRC, important species such as sea cucumbers and scallops have become extinct in traditional fishing grounds (ADB, 1995b). Mudflats that are seriously polluted have become unfit habitats for molluscs. Red tides and oil and industrial pollution are causing greater and increasingly frequent damage to coastal resources.

Other lesser known coastal resources are also under threat. For example, seagrass beds that serve as feeding areas for fish have been disturbed to the point where their species diversity has deteriorated. The value of these coastal resources to the community provides the greatest impetus for their conservation. In a recent study, a damage schedule measuring the relative importance of various resources was conducted for Bon Don Bay and Pak Phanang Bay, Thailand. For Ban Don Bay, the rankings, from most to least important, consisted of damage to mangroves, mudflats, shellfish breeding grounds, and fish breeding grounds. For Pak Phanang Bay, the rankings consisted of damage to sandy beaches, mangroves, seagrass beds, and coral reefs (Ratana, 1998).

Coral reefs in good condition yield 20 t or more of fish per 100 ha per year, and an average yield is estimated at 8 t/km^2 (ICLARM, 1999). Reef areas produce, inter alia, aquarium species and valuable live fish, mainly for East Asian markets and especially for restaurants in Hong Kong, China. The economic benefits from coral reefs are estimated at $375 billion per year (Bryant et al., 1998).

The biodiversity of coral reefs compared with other marine resource systems is likened to that of tropical rain forests in comparison with other forests. The range of species richness is from more than 2,000 species in the Indo-Pacific area to 200 species in the Atlantic. At present, about four fifths of all coral reefs are at risk and over half (56 percent) are at high risk levels (Bryant et al., 1998). Major threats to coral reefs are from overexploitation (36 percent), coastal development (30 percent), land-based pollution and soil erosion (22 percent), and marine pollution (22 percent). Southeast Asia contains about 30 percent of the world's coral reefs. Most of the areas with high species richness and high risk are (in order) in the Philippines, Indonesia, and Japan. In fact, all coral reefs in the Philippines are assessed to be at risk while in Indonesia about 80 percent are at risk.

Attempts are being made to protect aquatic resource systems through the establishment of Marine Protected Areas (MPAs), which number about 382 in Asia, excluding the Pacific, constituting about 29 percent of all subtidal MPAs in the world's 18 marine regions. The East Asian seas have MPAs in every biogeographic zone. Most of them are relatively small, and are threatened by various human activities. The management of these MPAs is constrained by inadequate funding and technical resources, shortages of trained staff and management information, inadequate commitment to law enforcement, the unsustainable use and overexploitation of resources, overlapping mandates, and a lack of coordination. Thus, despite the severity of the problems, relatively little protection has been accomplished.

Asia's marine areas vary greatly in their physical and environmental conditions, from mostly tropical in South and Southeast Asia, to subtropical to temperate and subpolar in East Asia (Devaraj and Vivekanandan, 1997). They cover areas of continuous and seasonal upwelling along the northwestern Indian Ocean, highly productive continental shelves along southwestern India, the South China Sea, the western Bering Sea, the area southeast of the Kamchatka Peninsula, and the Gulf of Thailand (part of the Sundaic platform), and thousands of islands with oceanic or near-oceanic features, such as Indonesia and the Philippines. When catches are expressed in terms of landings per unit shelf area, tropical oceans show the lowest levels of overall productivity. This is a reflection of the natural constraints imposed by a small supply of nutrients (FAO, 1997a).

The greatest diversity of fish species is found in the warm waters of the tropics, particularly the shallow inshore seas of the Indian Ocean and the western central Pacific Ocean. For example, there are 1,694 species of fish that reportedly inhabit Chinese waters. Of these, only 289 species are found in the temperate waters of the Bohai and Yellow seas, with the remaining 1,405 species being found in the tropical and subtropical waters of the East and South China seas (ADB, 1995b). In the temperate and subpolar areas in the northeast, the fishing industry is dominated by only a few species of fish. In warmer tropical waters, a great many species are fished. This is reflected in the wide variety of fishing gear and equipment used in these areas, and in the fact that most often there are many miscellaneous species caught at the same time as the targeted species.

Considerable uncertainties exist regarding the true potential of the oceans to supply fish, mainly due to a lack of data and incomplete stock assessments. This is particularly true in developing regions, but is also in part due to incomplete knowledge of fishery dynamics, prey-predator relationships, and environmental impact (e.g. the El Niño-Southern Oscillation phenomenon). Each of these may cause fluctuations in fish yields, making it difficult to estimate fishery potential.

However, even allowing for these uncertainties, there is growing evidence that the world's fishery resources in the major fishing grounds are being overexploited. Estimates of trophic levels for some 220 fish and invertebrate species or groups that are commonly landed (Pauly et al., 1998) have shown that there has been an increasing trend towards "fishing down the marine food web", possibly with severe implications for the marine ecosystem, especially prey-predator relationships. Fishing that removes the top carnivores in a food web may in the short term lead to higher production, because preyed-upon species would increase in the absence of their predators. In the long run, however, this practice could lead to widespread fishery collapses (ibid.). For example, the absence of a certain predator may increase the population of nonutilized competing predators (e.g., jellyfish in the Black Sea).

The lack of adequate stock assessments of common commercial fish species and the presence of large numbers of miscellaneous species in catches combine to make it very difficult to estimate the population size of any one species. This is a problem common to most fisheries throughout the world. Nevertheless, the traditional belief that the possibility of fishing a species to extinction was a remote one is now being challenged. The World Conservation Union (IUCN), for example, has listed over 100 marine species on its Red List of endangered species, including some tunas, sharks, and more than 30 species of seahorse (GRAIN, 1997).

The pressures from overfishing are particularly intense on larger fish species, which tend to have lower fertility levels and longer life spans, and are also usually commercially important species, e.g. the majority of groundfish, and tunas and other large pelagics. The smaller fish species tend to have high fertility levels and shorter life spans, and can therefore replenish their numbers relatively quickly, or are less intensively fished because of their lower commercial value. Recently, however, with the depletion of the stocks of the larger species and an increase in the demand for fishmeal for the animal feed industry, the smaller species are being fished more intensively.

In the Gulf of Thailand, the species composition of catches by trawlers has been changing away from long-lived and high-value species towards short-lived species of lesser value (Suraphol, 1997). An increasing proportion of catches from commercial fishing vessels consists of trash fish used for reduction. Similar trends of nonselective fishing are also becoming evident elsewhere, such as in India. Fishing effort continues to be high due to the need to make fishing operations cost effective and to make loan payments (Gopakumar, 1997). Fleet retirement is a costly option, both to governments and to fishers.

Major threats to the sustainability of fisheries and aquatic resources systems include mismanagement, and the lack of institutions that can ensure optimal exploitation and deal with multiple-use conflicts. For example, excess nutrients from pond aquaculture are discharged into the environment instead of being used to fertilize crops, and saltwater is brought into freshwater areas for shrimp farming. Cyanide and dynamite fishing in Philippine and Indonesian reefs continues to place the world's most productive reefs at risk. Heavy tourism and land and marine pollution exacerbate the situation. Dam construction disturbs the migratory pattern of fish. Soil erosion from deforestation and upland agriculture destroys coral reefs. Urban and industrial pollution creates costs for aquaculture that go uncompensated. Competition for resources between commercial and small-scale fishers often leads to mob protests and sometimes to violence. It is evident that governing institutions are under extreme stress and the current sectoral management approach, in which one agency is in charge of one resource without taking into account the interrelationships of resources in the same ecosystem, needs to be substantially reformed.

III SUSTAINABILITY OF ASIAN AGRICULTURE

T wo approaches can be used to review agricultural sustainability. The conventional, or spatial, approach tends to group the impact of agriculture on the environment into onsite, offsite, and global effects. Under this approach the emphasis is placed on the impact, making it useful for identifying affected groups or areas requiring mitigation and protection. In other words, it investigates where the symptoms are showing. Another approach is to look at the factors leading to unsustainable situations such that the causes of unsustainability can be pinpointed and tackled. The second approach, emphasized here, argues that technology management and government intervention are responsible for the growing symptoms of agricultural unsustainability.

THE RELATIONSHIP BETWEEN AGRICULTURE AND THE ENVIRONMENT

The impact of agriculture on the environment has both onsite and offsite aspects. Onsite, there may be direct negative impact on farm productivity or on the farmers directly involved in production. In this case, the actors bear the burden of their own activities. Offsite effects result in loss or damage that must be borne by those who are not party to production. In other words, offsite effects incur costs that are external to the actors. These external costs however, need not necessarily be physical.

Some empirical evidence seems to suggest that onsite effects are more significant than offsite effects because the loss is generally directly associated with highly productive areas. For example, the annual onsite cost of soil erosion in Java, Indonesia, in terms of losses in agricultural production has been estimated at $324 million, equivalent to 3 percent of the agricultural GDP of that island. The annual offsite costs were estimated at around $25 million to $91 million (Magrath and Arens, 1987).

Onsite Effects

When discussing onsite and offsite effects, it is important to distinguish between intensive and nonintensive agriculture. Onsite effects are largely related to mismanagement of intensive agriculture. These onsite effects were the biggest lessons learned from the green revolution. It was found that the HYV technology package required more complex management than was originally anticipated. Also, as mentioned earlier, the excessive use of agrochemicals created pest resistance and led to the emergence of new, more virulent pests and diseases. This situation has developed into a vicious circle of pest and insecticide overuse, resulting in both health and environmental problems (Box III.1). Other examples include concerns raised at the International Rice Research Institute (IRRI) in the early 1990s over yield declines, salinity buildup, increased incidence of soil toxicity and micronutrient deficiencies, hardpanning, changes in soil nitrogen supplies, and pest-related yield losses (Pingali and Rosegrant, 1993). Nonintensive agriculture, which generally needs relatively little input and which is more likely to be located in rainfed regions, has generated significantly fewer onsite effects. The environmental impact of nonintensified agriculture is related to expansion of agricultural areas or overexploitation of open-access resources, for example overgrazing.

Box III.1 Pesticide Use and its Impact on Human Health

Pesticide exposure can be direct, i.e. through contact on the job, or indirect, through residues in food or contaminated water and soil. Pesticide effects can be acute and immediate, or chronic. Many of those suffering from acute poisoning with such symptoms as headaches, nausea, or diarrhea are not hospitalized and their cases are not reported. Moreover, the effects from pesticides may be long term and cumulative. Not all the chronic effects are well understood. Acute poisoning can also cause health problems in later life. For example, people affected by acute organophosphate poisoning have been found to suffer neurological damage (WRI, 1998, p.44). Other chronic effects include dermatitis, immune system suppression, and male sterility from exposure to dibro-monochloropropane, which is used to control nematodes.

Offsite Effects

The offsite effects related to intensive agriculture are numerous. They tend to be the result of market and government intervention failures. A few examples are given here to illustrate the nature of their impact.

Water pollution in irrigated agriculture is often related to mismanagement and typically the overuse of chemicals. In the case of excessive use of nitrogen-based fertilizers, unused fertilizer may contaminate underground water supplies. Pesticide residues that contaminate agricultural products or the water supply could be harmful to human health, as well as to aquaculture. In the PRC, pesticide use in Zhujian, in the Yangtze basin, is one to three times greater than the national average. Yet only 20 percent of this is actually used by the plants, with the remainder being left in the soil or seeping into water sources (Zhang and Zhang, 1995).

Industrial livestock systems that develop near city centers generally have a high concentration of animals and have the potential to produce substantial organic discharges in excess of the carrying capacity of the surrounding environment. The intensive production of pigs in the PRC has caused animal waste pollution problems that now need close attention. Malaysia is also experiencing environmental problems arising from the pig sector. With proper management, these impacts can be alleviated and the environmental costs internalized and charged to consumers. For example, the Ponggol Pigwaste Plant in Singapore treats wastewater for recycled use at a cost of about 8-9 percent of the production cost of pork (Steinfeld, de Haan, and Blackburn, 1997). In Malaysia, aerobic waste treatment increases the cost of production of pork by 6 percent.

Without proper discharge management, under a high concentration and nutrient surplus system, the discharges could contain heavy metals that are harmful to animal and human health. Heavy metals such as copper, zinc, and cadmium are used as growth stimulants in some feeds. At present only the Organisation for Economic Co-operation and Development has regulations that aim to reduce the levels of heavy metals in feeds.

The offsite impact of intensive agriculture is not limited to pollution. Intensive aquaculture has considerable offsite impact through the destruction of mangroves as indicated earlier.

In less favorable environments where the green-revolution technology is not suitable, farmers tend to make up for low input use by area expansion. Offsite effects mostly involve the consequences of deforestation. Deforestation in upper watershed areas causes soil erosion and creates such offsite effects as sedimentation, which increases the costs to downstream industries. In the Philippines, the cost to the fishing and tourism sector from sedimentation due to uncontrolled logging in a 1,830-ha watershed was estimated at between $8 and 13 million (measured in terms of net present value) for fisheries and a loss of $9.2 million for tourism (Hodgson and Dixon, 1988, cited in Bann, 1998). Offsite costs resulting from the increased sediment load affecting irrigation systems, reservoirs, and harbors add another $90 million to the total.

Agriculture itself may be affected by pollution from other economic sectors. For example, air pollution from sulfur dioxide emission from electrical utilities may produce acid rain, which damages plants and animals as well as human health. In Thailand, some small-scale impact has been detected in sites near the Mae Moh power plant in Lampang. Industrial, land-based pollution and oil spills often threaten aquaculture. Natural resources and the environment are shared by many stakeholders; thus, multiple-use conflicts and external costs are unavoidable without appropriate intervention.

Climate Change

Global climate change, in the form of atmospheric warming, is occurring due to the release of greenhouse gases that accumulate in the atmosphere and increase the effect of radiation from the sun on the Earth. The changes in greenhouse gas concentrations are projected to lead to regional and global changes in climatic and related parameters such as temperature, precipitation, soil moisture, and sea level. However, the reliability of predictions surrounding the effect of climate change has yet to be proven. There are no hard facts concerning the result of increases in the concentration of greenhouse gases within the atmosphere, and no firm time scales exist.

Agriculture accounts for approximately one fifth of the annual increase in anthropogenic (human-made) greenhouse gas emissions (IPCC, 1996). The agricultural sector contributes to global warming through the emission of carbon dioxide, methane, and nitrous oxide.

Methane and nitrous oxide from agricultural sources contribute about 50 and 70 percent, respectively, of global anthropogenic emissions of these gases. Their main sources are flooded rice cultivation, the use of nitrogen fertilizers, improper soil management, land conversion, biomass burning, and livestock production, including the associated manure management. It has been claimed that the livestock industry contributes between 5 and 10 percent of the overall contribution to global warming.

Deforestation and the burning of agricultural crop wastes or rice stubble remain major sources of carbon emissions. When natural systems are converted into agricultural land, a large proportion of the soil carbon can be lost because plants and dead organic matter are removed. This process contributes approximately one third of total global carbon dioxide emissions. To a lesser extent, carbon dioxide is released from the use of fossil fuels in agricultural production, and from livestock production. High-intensity animal production has become the biggest consumer of fossil fuel energy in modern agriculture (IPPC, 1996).

Within the agricultural sector, methane is the most significant greenhouse gas released. Most of the releases come from rice fields (91 percent), the remainder being from animal husbandry (7 percent), and the burning of agricultural wastes (2 percent). The quantification of emissions from rice fields has proven difficult because the emissions vary with the amount of land in cultivation and also depend on fertilization practice, water management, density of the rice plants, and other agricultural practices. The PRC is a very large source of methane in comparison with other Asian countries (Table III.1).

Livestock and associated manure management contribute 16 percent of the total annual production of methane. These emissions are a direct result of consumption by cattle and buffaloes of large amounts of fibrous grasses that cannot be used as human food or as feed for pigs and poultry. Cattle and buffaloes account for about 80 percent of global annual methane emissions from domestic livestock.

The main source of nitrous oxide released from agriculture arises from the excessive use of nitrogen-based fertilizers, legume cropping, and animal wastes. The flux of nitrogen depends on the microbial activity of the soil. For example, wet rice absorbs only one third of the nitrogen in the fertilizers and upland crops about half. The rest is denitrified and diffused into the atmosphere, contributing to global warming. However,

Table III.1: Methane Emissions from Livestock and Agricultural Sources
in Selected Asian Countries, 1990

	Methane (t'000)	
	Livestock	Other Agriculture[a]
Bangladesh	520	473
China, People's Rep. of	8,940	18,400
Indonesia	864	2,039
Japan[b]	520	276
Kazakhstan	939	
Mongolia	301	
Nepal	370	542
Philippines	315	559

[a] including flooded rice fields
[b] 1994

Source: WRI (1998).

the amount of nitrous oxide emitted is much lower in volume
than the amount of methane.

The aggregate global effect of climate change on
agricultural production is likely to be small to moderate.
However, climate change will have significant regional impact
on agricultural yields. Crop yields and changes in productivity
will vary considerably across regions and probably result in a
slight overall decrease in world cereal grain productivity.
Decreases in productivity would be most likely in regions that
already experience food shortfalls.

The effects of climate change will also differ across Asia.
The changing temperature as well as changes in rainfall patterns
and the accompanying increase in projected levels of carbon
dioxide will have important effects, especially in tropical regions.
It is expected that food productivity (especially crop productivity)
will alter due to these changes in climate, and due to weather
events and changes in distribution of pests and diseases. Land
areas suitable for the cultivation of key staple crops could undergo
geographic shifts in response to climate change.

Vulnerability to climate change depends not only on
physical and biological response but also on socioeconomic
characteristics. Low-income populations depending on isolated

agricultural systems are particularly vulnerable to hunger and severe hardship. In these areas, where populations are already barely food sufficient, even the slightest decline in yields could be very harmful. The most negative effects foreseen are in dry land areas at lower latitudes, in arid and semi-arid areas, especially those reliant on rainfed agriculture. Many of these at-risk populations are located in South and Southeast Asia.

Impact on rice yields in South and Southeast Asia is likely to vary greatly (Matthews et al., 1994a, 1994b). Several major studies have been conducted of countries in East Asia, including the PRC, the Republic of Korea, the Democratic People's Republic of Korea, and Japan (IPCC, 1996). While large changes were predicted for the PRC, the studies concluded that to a certain extent, warming would be beneficial, with yields increasing due to a diversification of cropping systems. Studies for Japan have shown that the positive effects of carbon dioxide on rice yields would generally more than offset any negative climatic effects.

Climate change could influence food production adversely in three ways: geographical shifts and yield changes in agriculture, reduction in the quantity of water available for irrigation, and loss of land through a rise in sea level, which would also cause salinization of coastal land. Geographic limits and yields of different crops may be altered by changes in precipitation, temperature, cloud cover, and soil moisture as well as by increases in carbon dioxide concentration. High temperatures and diminished rainfall could reduce soil moisture in many areas, particularly in some tropical and midcontinental regions, reducing the water available for irrigation, and impairing crop growth in nonirrigated regions.

Changes in soils, e.g. the loss of soil organic matter, the leaching of soil nutrients, salinization, and erosion, are likely consequences of climate change in some climatic zones. The risk of losses due to weeds, insects, and diseases is likely to increase. The range of many insects will expand or change, and new combinations of pests and diseases may emerge as natural ecosystems respond to altered temperature and precipitation profiles. The effects of climate change on pests may add to the effect of other factors, such as the overuse of pesticides and loss of biodiversity that already contribute to pest and disease outbreaks.

Agriculture in low-lying coastal areas or adjacent to river deltas may be affected by a rise in sea level. Flooding will probably become a significant problem in some already flood-prone regions of Asia such as the PRC and more southern parts of East Asia. The summer monsoon is predicted to become stronger and move northwestward. However, the resulting increased rain could be beneficial to some areas.

Climate change could affect both livestock and dairy production. The pattern of animal husbandry may be affected by alterations in climate and cropping patterns, as may the ranges of disease vectors. In warm regions, higher temperatures would likely result in a decline in dairy production, reduced animal weight gain and reproduction, and lower feed-conversion efficiency. More mixed impacts are predicted for cooler regions. If the length and intensity of cold periods in temperate areas are reduced by warming, feed requirements may be reduced, the survival of young animals enhanced, and energy costs for the heating of animal quarters reduced.

Climate change would also affect livestock through its impact on disease. The incidence of diseases of livestock and other animals is likely to be affected by climate change, since most diseases are transmitted by vectors such as ticks and flies, the development stages of which are often heavily dependent on temperature. Sheep, goat, cattle, and horses are also vulnerable to an extensive range of nematode worm infections, most of which have development stages that are influenced by climatic conditions.

In general, intensely managed livestock systems have a greater potential for adapting to climate change than do crop systems. Adaptation may be more problematic in pastoral systems where production is very sensitive to climate change; technology changes introduce new risks, and the rate of technology adoption is slow. Livestock production may also be affected by changes in grain prices, and rangeland and pasture productivity.

In developing countries, livestock are better able to survive severe weather events such as drought than are crops, and are therefore a better option in terms of income protection and food security (Abel and Levin, 1981).

Various types and levels of technological and socioeconomic adaptations to climate change are possible. The extent of adaptation depends on the affordability of such measures, particularly in developing countries. Recent national studies show that the increased costs of agricultural production under climate change scenarios would be a serious economic burden for some developing countries. Other important factors are access to know-how and technology, the rate of climate change, and biophysical constraints such as water availability, soil characteristics, and crop genetics. Improved land-use practices may help to mitigate greenhouse gas emissions. Some structural changes in agricultural production could also be beneficial and may reduce the necessity for soil disturbances, e.g. switching from rice to other crops such as sugar. However, rice will remain an important food crop in Asia.

Significant decreases in methane emission from agriculture could be achieved through better management of rice fields and by reduced biomass burning. A reduction in methane emission could be achieved by a shift from the use of organic manure to mineral fertilizers (Wasson, Moya, and Lantin, 1998), a shift from traditional to high-yielding crop varieties, the intermittent drying of soils, and zero tillage and mulching. Irrigated rice has been found to produce more methane than deepwater rice (Charoensilp, Promnart, and Charoendham, 1998). The appropriate application of chemical fertilizers, changes in cultivation practices (such as a shift from transplanting to direct seeding), and appropriate water management can also contribute to reducing methane emission. These combined practices could reduce methane emission from agriculture by 15 to 56 percent. Energy use by the agricultural sector has decreased greatly since the 1970s. However, fossil fuel use by agriculture and the resulting carbon dioxide emissions could be further reduced by such actions as minimum tillage, irrigation scheduling, the solar drying of crops, and improved fertilizer management.

Additional methane reduction is possible by improved nutrition for ruminant animals and modifying the treatment and management of animal wastes. The shift to monogastric animals such as pigs and poultry results in a lower level of methane emission because these animals have different feed requirements from cattle. However, opportunities for further reducing methane emission from intensively managed cattle are somewhat limited because the methane production per unit of cattle feed is small and cattle are already being given a high-quality diet. Nitrous oxide emissions could also be decreased through better treatment and management of animal wastes.

It is important to note the role of forests and vegetation as greenhouse gas sources and sinks. The emission of carbon dioxide is only one part of the carbon cycle. The assimilation of carbon dioxide also occurs where vegetation binds carbon into biomass. Carbon storage in the soil is important and dependent on the type of vegetation. Vegetation and soil from unmanaged forests hold 20 to 100 times more carbon per unit area than does agricultural land. Deforestation and land-use changes have diminished the global storage of carbon as well as the land's capacity to bind carbon dioxide.

Although opportunities to reduce the emission of greenhouse gases exist, the problem is that options usually require a trade-off between productivity and emission. It is important to investigate these trade-offs so that appropriate policies and incentives can be designed.

In conclusion, although global warming is expected to have some impact on tropical agriculture, especially in arid and low-lying areas, the specific locations and timing of the projected impact remain uncertain.

FACTORS DETERMINING AGRICULTURAL SUSTAINABILITY

Sustained increases in agricultural production depend on the availability and the quality of natural resources and the way humans interact with nature in the production process. The interaction between humans and nature depends on the availability of resources, crop choice and technology, incentive systems, and the rules and regulations that govern the use of resources. Central to this interaction are two important factors that determine agricultural sustainability: technology management and government intervention.

Technology facilitates the exploitation of nature. It provides the methods used in interacting with and making an impact on nature. It has also been a major instrument in saving natural resources, maintaining and extending nature's carrying capacity, mitigating negative effects, and enhancing positive impact on the environment. Technology designed to meet production or extraction objectives can be environmentally neutral, enhancing, or destructive. Technology can also produce second-generation effects that induce declines in productivity and undermine long-term sustainability. The creation and adoption of technology depend on its profitability, which in turn is influenced by the prevailing incentive system and the rules and regulations related to production decisions.

Generally speaking, developmental and political objectives and policies in Asia have created the incentive systems in the various economies. In order to meet food security and income objectives, maximization of output has often been the overriding goal. In low-income countries and production units, short-term needs tend to take precedence over long-term gains. Governments want to be perceived as givers rather than takers. Therefore, subsidies are readily extended, while taxes or levies are imposed only reluctantly. Technical or engineering solutions such as technological innovation and infrastructure development are easier to administer than rules and regulations. Administrative responsibilities are often sectoralized such that benefits and power can be shared. The status quo is preferred to change as the latter implies winners and losers. For example, using technology that saves land is preferred to land reform. Rewards are bestowed from the top and not determined by the people whom the government machinery is created to serve.

The above modus operandi works well with the green-revolution technology package, at least in the initial phase of technology transfer. The green-revolution technology is scale neutral and does not require institutional reform. It requires investment in irrigation and transportation networks. This genetics-based technology is powerful enough to propel growth in areas where the environment is favorable. However, the same technology will not be able to maintain such miracles in the long term; the mode of operation needs to be adjusted to meet changing needs. Reasons for this are put forward below in terms of technology management and government intervention.

Technology Management Issues

The environmental impact of agriculture varies with the type and the level of intensification. In areas where the environment is favorable, agriculture is usually intensive and high-input technology is often used. Moreover, good water availability makes the use of chemicals more worthwhile. Higher output from the technology in turn encourages the overuse of chemicals, resulting in both onsite and offsite environmental effects that undermine the sustainability of the agricultural sector.

Sustainability of the Intensive Monocropping System

Much of the concern regarding agricultural sustainability is related to yield declines in the intensive monocropping system. The last 20 years have seen the emergence of many site-specific problems in these systems throughout Asia, e.g. boron toxicity and zinc deficiency in rice-rice-rice cropping at IRRI, and boron deficiency in wheat in the rice-wheat cropping system in Bangladesh, PRC, India, and Nepal. The development of such problems is not surprising in cropping systems that have become increasingly intensive. With two and sometimes more crops being grown in succession on the same land and in the same year, major biological, chemical, and physical changes have taken place in the soil.

Boron deficiency occurred in HYV wheat in Bangladesh, northeastern India, and in Nepal because the introduced varieties were not accompanied by screening for boron efficiency. Again, this points to a lack of awareness of site-specific problems. The experience of the PRC shows that many problems related to intensive cropping are site specific and can be managed through strong and responsive R&D in crop management.

Iron toxicity from continuous flooding was identified in Indonesia, Malaysia, Philippines, and Sri Lanka. However, it happened before the widespread adoption of HYVs (Tanaka and Yoshida, 1970). Reported incidences of zinc deficiency in rice also preceded the release and adoption of HYVs. Increased incidence of micronutrient deficiencies could also simply be the result of improved diagnostic capabilities, allowing the identification of previously undetected problems.

Hardpanning or subsoil compaction is a problem that occurs when an upland crop is grown on land previously used to cultivate rice. Recent experiences have demonstrated that there is no difference in yield when the soil is subject to zero tillage or multiple tillage (Hobbs, Sayre, and Ortiz-Monsterio, 1998). In the PRC, reduced soil porosity was solved by tilling the soil once every three years (Wang and Guo, 1994) or by deep ripping.

Changes in supply capacity of the soil were observed during long-term experiments on rice (Cassman, Peng, and Dobermann, 1997), where nitrogen (N) fertilizer levels had to be increased from 140 to 200 kg/ha. Although observed in a research environment, this problem has yet to be encountered in fields under cultivation.

The above complications indicate that genetic improvements alone are not the answer to the food production problem. Second-generation problems do arise, but with good local R&D and crop management with special emphasis on the sustainable management of soil and water, many of these problems can be solved and productivity gains can be maintained. The sustainability of intensive cropping systems will depend on the capacity of the local R&D system for timely identification of the problems and provision of solutions to these problems.

Pest Control

Asia accounts for 16 percent of global pesticide sales (WRI, 1998), and developing countries overall account for about half of all pesticides used (Alexandratos, 1995). The impact of pesticide use on human health is believed to be great. Although the total number of persons affected is uncertain, it is thought to be between 50 and 100 million (WRI, 1998, p.44). Agricultural intensification near the Aral Sea in Uzbekistan has been blamed for pesticide-related illnesses, and impact on farmers' health has been reported in the Philippines (Loevinsohn, 1987; Rola and Pingali, 1993).

Another effect of chemical overuse is the development of pest resistance, leading to an even greater use of chemicals. In this regard, the brown plant-hopper (BPH) epidemics can be singled out as an unexpected side effect of the green revolution. The BPH was a minor rice pest in Asia before the green revolution. The nonselective use of chemicals that accompanied the technology package destroyed the predators of the BHP and transformed it into a major pest. In response to the outbreaks, more pesticides were used. The insect finally evolved into more virulent biotypes that can break down the resistance of some high-yielding rice varieties. Moreover, pesticide overuse also created chemical resistance in the pests.

The BPH epidemics took their heaviest toll in Indonesia, where chemical inputs were subsidized under a food-sufficiency program. In 1977, a BPH epidemic caused Indonesia to lose more than one million t of rice worth more than $100 million. A 1986 BPH recurrence caused even greater damage, estimated at $400 million.

More recently, resistance to the selective herbicide isoproturon, used for controlling little-seed canary grass, in the rice-wheat system in northwestern India, has affected almost one million ha of wheat (Malik and Singh, 1994; Malik, Gill, and Hobbs, 1998).

The collection and accumulation of genes necessary to build up resistance to pests takes a decade or more. Thus, scientists found that a breeding strategy would not be sufficient or timely enough to cope with the problem. Finally, it was decided that integrated pest management (IPM) needed to be adopted. IPM is defined by FAO in the "International Code of Conduct on the Distribution and Use of Pesticides" (Article 2) as "a pest management system that, in the context of the associated environment and the population dynamics of the pest species, utilizes all suitable techniques and methods in as compatible a manner as possible and maintains the pest populations at levels below those causing economically unacceptable damage or loss". It is a holistic approach to pest control where a combination of various control methods are used, including selective chemicals as well as natural predators and parasites. IPM has finally brought BPH under control, particularly in Indonesia (Box III.2).

Indonesia is not the only country that can claim success in reducing pesticide use. The PRC has also been reducing pesticide use since 1982 owing to improved pest management and improved quality of pesticides (Fan, 1997). However, crops not associated with the green revolution, e.g. horticultural crops and cotton, are much more intensive in their use of pesticides. In India, half of all pesticides used are for cotton crops (Paroda and Chadha, 1996), which account for only 4 percent of total crop area.

Nutritional Imbalance

Fertilizer use is often very high under intensive cropping systems such as in the PRC's northern plains, middle and lower Yangtze River valleys, and Sichuan basin, and in Haryana and Punjab, India. In horticulture, annual organic fertilizer use

---

Box III.2   IPM Farmers

IPM requires some basic knowledge of insect ecology and the toxicology of insecticides. This knowledge was at first considered too difficult for farmers in most Asian countries to learn, because most of them have little formal education. The IPM strategy requires farmers to be para-taxonomists and ecologists. Successful IPM farmers need to recognize and monitor their crops' natural enemies and to take appropriate steps. Rules exist to help determine appropriate actions, but no one single formula for arriving at a remedy exists.

Indonesian farmers were able to prove that when the proper policies were implemented they could overcome the BPH plague and achieve a win-win solution, with both an increase in rice production and a decline in pesticide use. The IPM program was launched in 1986 and by 1989 the necessary associated policy of removing pesticide subsidies was instituted. About $1,200 a year per farm was saved by reduced pesticide use, a total estimated benefit of $1 billion. By 1993, Indonesia had 250,000 IPM farmers.

Large numbers of IPM farmers are also being trained in the PRC, India, and the Philippines, (Oka, 1996). In all countries where IPM has been adopted, rice yields are greater than under conventional methods.

---

easily exceeds 2,000 kg/ha. Two interrelated problems have arisen as a result of this high input of fertilizer:nutrient imbalance in the soil and offsite effects from the overuse of chemicals.

Nutritional imbalances occur when the amounts of various nutrient elements required by crop plants are not matched by those supplied by the soil and fertilizer. Among the 15 or 16 mineral elements essential to plant growth, those most often deficient in Asian soils, as elsewhere, are nitrogen (N), phosphorus (P), and potassium (K). Fertilizers used in crop production in Asia almost always contain N, and increasingly also P and K.

As most of the fertilizer now applied is N-based, it is inevitable that deficiency of other nutrients will become the next limiting factor to yield growth. About two thirds of agricultural land in the PRC and almost half of the districts of India are considered to be affected by phosphorus deficiency (Stone, 1986; Desai and Ghandhi, 1989). In the PRC, it has been estimated that a yield increase of up to 18 percent (almost equivalent to the gain made by their famous hybrid-rice technology) could be obtained by improving nutrient management (Lin and Shen, 1994). Measurements of plant nutrient status in farmers' fields indicate that the nutritional balance is often poor. Onfarm studies have shown that Asian rice farmers do not often apply N, P, and K fertilizers in amounts that correspond with the soil's capacity to supply these nutrients (Cassman, Peng, and Dobermann, 1997). In addition, new diseases are now increasingly identified as being associated with nutritional imbalances (Cassman, Peng, and Dobermann, 1997). Considerable productivity gains as well as ecological benefits might be expected from improving the nutritional balance in fertilizers. So far, phosphorus, potassium, sulfur, boron, manganese, copper, and zinc deficiencies have been identified as factors that can limit crop yields in various locations in Asia.

While fertilizers are underutilized in most parts of Asia, overfertilization occurs in some favorable environments, especially in intensive vegetable production (Morris, 1997). In addition to the cost, there are ecological and health consequences of excessive fertilization. Very high levels of phosphorus (over 1,000 parts per million (ppm), where a "good" soil contains about 15 ppm), have been found under vegetables in different locations in Asia. This shows a lack of knowledge on nutrient management. Unused nitrogen from fertilizers ends up as nitrates in underground water, or in streams where intensive vegetable crops are grown in the highlands (e.g. the Philippines, Thailand). A survey of 3,000 dug wells in Indian villages showed that about 20 percent of them contained nitrate levels in excess of the WHO limit (Handa, 1983).

The decline in rice yields in IRRI's long-term experiments in the 1970s was found to be due to boron toxicity, resulting from irrigation with well water containing high levels of boron, combined with zinc deficiency (Flinn et al., 1980). Marginal grain-to-nutrient ratios from use of these micronutrients can be expected to be much higher than that currently gained from nitrogen alone. The difficulty lies in determining where and how the gains can actually be achieved on farms. As most of the fertilizers used in Asia are nitrogen based, and 40 to 66 percent of the nitrogen now applied to rice and about half that applied to other crops are actually wasted, some significant gains should also be realized from improving the management of nitrogen.

Waterlogging and Salinity

Waterlogging and salinity, although often topics of discussion in literature related to land degradation, are a result of the mismanagement of water. Waterlogging and salinity problems occur in areas where excessive irrigation induces salt build-up through capillary action.

Global estimates of the significance of the problem differ widely, between 18 million ha (Postel, 1992, cited in Crosson, 1994) and 43.5 million ha (Dregne-Chou, 1992, cited in Crosson, 1994). Some 10 to 15 percent of the irrigated land in developing countries is somewhat degraded through waterlogging and salinization (FAO, 1995a), and "waterlogging and salinization have sapped the productivity of nearly 5 percent of the world's (250 million ha) irrigated land" (FAO, 1995a). Salinity affects 11 percent of the irrigated land in India, 21 percent in Pakistan, and 23 percent in the PRC (FAO, 1995a).

Using the Postel and FAO estimates, Crosson (1994) calculated the rate of annual increase in salinization to be 2.3 percent, and constructed three supply scenarios based on different assumptions that gave comparative yields from affected and unaffected land. He concluded that the impact on global output by 2030 would be a loss of 3 to 16 percent of production.

It should be noted, however, that it is difficult to differentiate between "intensification-induced" and naturally occurring salinity problems. For example, salinity in the lower Mekong Delta in Viet Nam, caused by the low river flow in the dry season, affects 1.7 million ha of agricultural land as well as other economic activities. The problem is so severe that controls were put on water use for agriculture to maintain a critical flow (Mie Xie, 1996). Without precautionary measures, the total area affected could reach 2.2 million ha.

Waterlogging and salinity problems are another consequence of the singularly crop-centered approach of the green revolution, in that it has neglected aspects of crop management not based on fertilization and irrigation. An integrated approach is needed for salinity-prone areas (Qureshi and Barret-Lennard, 1998), for example the introduction of salt-tolerant wheat, water-table management by planting deep-rooted trees for drawdown, and the planting of halophytes such as salt bushes. These problems indicate the need to intensify crop management research, which has lagged behind breeding research, for example on rice at IRRI.

Some of the region's salt-affected irrigated lands are, however, still not free from the risk of further degradation. This is caused by two basic but related problems: (a) the use of salt-laden irrigation water, and (b) the disposal of the extra salt. Pakistan's irrigation system, for example, adds 60–65 million t of salt as saline water to the underground supply annually, 35–40 million t of salt as canal water, and 20 million t of salt from "fresh" (i.e. better quality) underground water onto irrigated lands. The major saline-effluent disposal projects now under development are expected to carry only a fraction of this salt out to sea (Qureshi and Barrett-Lennard, 1998). Surface soils that were only moderately saline or salt-free have become severely salinized. In addition to increasing the accumulation of salt in the soil profile, using poor quality irrigation water further degrades the soil by destroying its physical structure, making it impermeable to water. Leaching salt from such soils is difficult, and crop growth is then adversely affected by waterlogging as well as by the salt. Some 2–3 million ha in Pakistan have already been reported as having suffered further degradation in this way (Rafiq, 1990).

Genetic Erosion

The loss of genetic diversity following the widespread adoption of HYVs and other MVs in Asia has raised two concerns. The first is related to the fear that traditional varieties will be lost as farmers narrow their crop choice; over three quarters of wet riceland in Asia is now planted with MVs. This concern is currently being addressed by the storage of traditional varieties in international gene banks. The second concern is related to the increased risk of pests and pathogens associated with large-scale production of genetically uniform varieties, a risk that was demonstrated by the 15 percent yield loss in maize in the US in 1970 due to the southern corn leaf blight.

Evidence in Asia tends to suggest that the variability of output, as measured by the coefficient of variation, has decreased. For instance, in India the coefficient of variation of wheat yields between the decade before the green revolution and 1976 to 1986 decreased from 17 to 7 percent (Singh and Byerlee, 1990). Neither maize nor wheat has suffered major outbreaks of pests or pathogens since the green revolution. This remarkable achievement is attributable to the fact that plant breeders have been able to develop new varieties at a rapid pace, especially varieties with a strong resistance to rust pathogens.

The story of rice is quite different, with major outbreaks of disease and pests having occurred in large areas where a single variety is cultivated. Examples of this are the BPH epidemics mentioned earlier and the blight that struck the rice variety RD6 in Thailand.

At present, the coverage of major staple crops by germplasm held in international gene banks provides sufficient guarantee against genetic erosion. In 1992, rice, wheat, maize, and soybean accessions totaled 250,000, 410,000, 100,000, and 100,000, respectively (Chang, 1992, cited in Evenson, 1996). Cultivars of wheat and rice then uncollected constituted about 10 and 5 percent, respectively. There is, however, concern about continuity of international funding to sustain these gene banks and to improve collections from the wild such that more systematic and useful information can be provided.

The lesson to be learned from onsite effects due to the intensive cropping systems, pest control methods, and from offsite effects or externalities, is that for sustainable growth of a crop-centered technology, more emphasis is needed on soil, water, crop, and genetic management. The current capacity of management in these areas lags behind the technology, both at the national and the international level.

Failures in Government Intervention

For the purposes of policy and planning, it is useful to categorize government interventions into those at the project, sectoral-policy, and national-policy levels. In this volume, only failures at the first two levels are discussed because national policies are generally designed to serve much broader economic objectives such as full employment and economic stability. More information on the impact of national policies on rural Asia is available in a companion volume (Rosegrant and Hazell, 1999). However, it is important to note the impact that national policy variables could have on the sustainability of agriculture. For example, the higher the interest rates, the more difficult it may be for economic agents to invest for the long term, including investing for gains from conservation practices. The more undervalued the exchange rate, the greater may be the exploitation of natural resources for exports. For sectoral policies, emphasis is given here on discussion of natural resource sector policies.

Government failures mainly come from one of four sources: intervening in a market that is functioning well, neglecting to correct for market failures, inefficient provision of public goods, and inadequate consideration of trade-offs and opportunity costs (Panayotou, 1993). In the first instance, government policies may distort prices, generally by overpricing output through price supports and guarantees, and by underpricing inputs through subsidies. This has been very prevalent in the agricultural sector. Second, market failures or functions that the market cannot perform, for example to achieve allocation efficiency, are not addressed. Offsite or external effects (such as pollution and deforestation) are not effectively handled. Natural resources are left under an open-access regime and are wastefully exploited. Third, a government may overextend itself by pursuing activities best accomplished by other institutions and agencies. For example, central governments often attempt to provide local public goods directly, rather than create a situation where the provision becomes possible through other agents, in this case local government and communities. Finally, inadequate consideration may be given to trade-offs and opportunity costs, especially when the costs are not readily measurable or not expressed in monetary terms.

Trade-off situations have occurred when there is no win-win solution or when two or more objectives cannot be simultaneously fulfilled. Then there will have to be losers as well as winners. In such cases, the option with the highest benefit is normally selected, provided that the cost-benefit analysis has been thorough and comprehensive. In the area of natural resources and environment, most environmental damage is not immediately obvious, but attempts must be made to quantify the damage.

Trade-offs may occur at the farm, sector, national, or international level. At the farm level, long-term income and sustainability may have to be forfeited for short-term gain owing to resource constraints or immediate hunger. For example, in the uplands, soil conservation may not be adopted because the additional labor required competes for labor to be used in food collection. At the sector level, a new dam that will enhance agricultural productivity may be constructed at the expense of severe biodiversity loss or at the expense of an existing dam that relies on the same inflow. At the national level, trade-offs may be in the form of a choice between accelerated growth in the short run and sustained growth in the longer term.

Project-Level Failures

Project-level failures arise from a lack or neglect of intense and careful information gathering and exchange between the many and diverse stakeholders, especially those at the grassroots level. Projects are frequently designed and undertaken with limited scientific data, without respect for social practices and norms, and without taking advantage of local knowledge and wisdom. Project selection criteria are not based on rigorous and thorough analysis of cost and benefits and often exclude the cost of environmental damage and external effects altogether. Trade-offs and opportunity costs are not given careful consideration and therefore careful accounting is not done for them. In addition, the project time span is often too short for the achievement of project objectives and can indeed influence objectives to be short term.

The establishment of the cattle industry in developing countries is a good example of how the role of information is often underestimated in public investment projects. In Thailand, the Government embarked on a project to promote foreign breeds, which were later dubbed "plastic cows" because farmers were simply not told of the nutritional levels required to achieve full reproductive efficiency. In Nepal, a similar mistake was made with the use of continuous backcrossing with Holstein Friesian cows. In that instance, the situation was even worse because the cow is sacred according to religious tradition and cannot be destroyed even if it does not calve every year and therefore fails to provide milk.

Many failures have occurred with project-level irrigation projects. In northeastern Thailand, the Nam Bor reservoir had to be drained and abandoned after it was completed because it was built on a dome of rock salt and had led to salinity build up. The most classic irrigation failure was the attempt to convert first-class grazing land in Uzbekistan and Kazakhstan in Central Asia into second-class irrigated land. This project resulted in the immense environmental disaster of the drying of the Aral Sea (Turner II and Benjamin, 1994). The Syr Darya and Amu Darya rivers, which flow into the Aral Sea, are heavily drained to irrigate cotton. From the Amu Darya alone, 14 cubic kilometers of water, about 90 percent of all annual renewable water resources in Uzbekistan, are drained into the Kakarum Desert. Despite the desiccation of the sea, the project is overextended and fails to deliver water efficiently, resulting in a loss of up to 70 percent of the water during delivery. Moreover, salinity build up neutralizes much of the benefits of irrigation.

Increasingly, conflicts between stakeholders have delayed and prevented projects because opportunity costs and trade-offs are not well understood and accepted by government agencies and thus are not skillfully handled. Compensation rules allowed under government regulations only compensate for the loss of nonmovable assets such as trees and houses, and not for opportunity costs such as the income forgone from traditional fisheries, or the harvest of nontimber forest products.

Many countries now require an environmental impact assessment (EIA) before granting project approval. This exercise tends to be done in order to meet requirements of international lending agencies rather than being a serious effort to mitigate environmental impact. Environmental impact is summarily appraised although in many cases, year-round observations of potential impact are necessary. In some countries, environmental experts who are invited on a voluntary basis to consider the EIA lack adequate support to investigate the project thoroughly. In other countries, experts within the public sector do not exist. Some consultant agencies do not have adequate skills in handling environmental issues and often mitigation costs are not included in the final cost-benefit calculations of a project. Public participation procedures do not exist in many countries. The procedure of approval and information about projects are not transparent and are sometimes withheld from the public. Public hearings are usually not required (Mingsarn et al., 1998).

Project failures also result from the fact that benefits and costs are narrowly defined without consideration being given to the resource system. For example, investment in water resource development is often project based rather than basin based. Projects are often considered for individual merits without a careful ranking of benefits and costs of related projects in the same resource system.Projects are sometimes implemented despite very low rates of return and high external costs.

Agricultural funding from both internal and external sources has been project oriented. While this may continue, attempts should be made to convince governments to engage in policy and institutional reform. For the next decade, policy and institutional reform would probably release more productive resources and enhance more output than investment. In the future, project-based development should be a component of policy reforms, which are most urgently needed in the areas of natural resource management, R&D, and extension.

Finally, the current management systems that concentrate on legal instruments and command-and-control regimes as the main mechanisms for resource management, have missed out on other management opportunities offered by market-based and fiscal instruments such as taxes, charges, and incentives. The legal instruments also fail to acknowledge specific regional variations and limit the ability of the State to take advantage of local knowledge and initiatives. The devolution of R&D for crop management and for the management of some critical resources is the first step to sustained production increases.

Sectoral Policy Failures

Despite many success stories, price distortions from government support and subsidy programs can still be found in Asia. For example, export taxes on agricultural products and import duties distort the allocation of resources for crops, which in turn has impact on the environment. The impact differs from country to country, depending on the type of crop and also on whether the crop is imported or exported. Such impact is therefore an empirical issue. In Sri Lanka, it has been shown that liberalizing trade could have had a positive effect on the environment (Coxhead and Jayasuriya, 1992). The result has been reduced soil erosion as well as a positive income effect due to the plantation crops (namely tea and rubber that are subject to heavy direct and indirect tax) being more environmentally friendly than food crops (Chisholm, Ekanyake, and Jayasuriya, 1997). In the Philippines, export taxes and an overvalued exchange rate encouraged the production of annual crops such as potato, cabbage, and garlic at the expense of more environmentally friendly export crops like coffee, cocoa, and rubber (Coxhead and Rola, 1998). Agricultural research policy in the Philippines also favors vegetables, legumes, and root crops, which are relatively less environmentally friendly than the woody perennial species. Policies aimed at increasing food security have favored the expansion of corn into the marginal uplands in the watershed in Lantapan.

Food security policy in itself is often a source of agricultural unsustainability. In Bangladesh, many policies have been geared towards rice production and against wheat, although there are some important ecological niches, such as higher elevation and lighter soils, where wheat is more favorable than *boro* rice (winter or dry season rice) (Morris, Singh, and Pal, 1998). Wheat is much less water intensive, making it suitable for areas without access to irrigation. However, the price of wheat is held below import prices owing to food aid and subsidies.

Input subsidies have encouraged excessive chemical use, resulting in pest epidemics and heavy yield losses. There is now an immense literature showing that removing policy distortions will provide win-win solutions. The Indonesian Government, for example, is acclaimed for its action in this direction by removing pesticide subsidies, thus encouraging the use of integrated pest management (Box III.2). Win-win solutions can be expected in India where the removal of fertilizer subsidies would provide greater incentives for the use of manure in soil management, which in turn would raise the demand for labor, especially for women (World Bank, 1996). However, in this particular case in India, the shift in the use of manure as a source of fuel to fertilizer may have to be set against a negative impact on forests.

As food security in Asia has increased, the agricultural policies mentioned above have lost much of the justification for their existence. In addition, the cost of subsidies has become overwhelming. Most of these high-cost policies have been maintained for political, not economic, reasons. Issues related to policy failures today have shifted from agricultural to natural resources policies.

Policy failures concerning natural resources are often the result of open-access policies. A classic example is oceanic fisheries resources: any fisher who is willing to invest in a boat can harvest as much as his boat will carry. The best known example of environmental degradation under an open-access regime was brought to the attention of scientists and the public when Garett Hardin wrote about the "tragedy of the commons". Hardin (1968) depicted a pasture that was open to all. Overgrazing and a degraded pastureland would be the inevitable outcome because each user would tend to raise as many animals as possible in order to maximize private gains. Hardin's "commons" represent an open-access resource, in which each individual tries to maximize private gain by converting public into private property. Consequently, the first-come-first-served situation that exists under open-access regimes tends to encourage the wasteful use of resources.

Land Policy and Institutions

In most countries in Asia, land tenure security, either in the form of ownership or long-term lease, is increasingly recognized as an important incentive for attracting investment in land to improve its productivity. In fact, land is the most privatized natural resource. Efforts are now underway in many countries to define use and ownership rights and to provide the corresponding recognition. The process is often slow, however, and in many instances tarnished with corruption.

The lack of well-defined rights to land use can lead to substantial environmental degradation. According to a detailed analysis of the PRC's pastoral region in the northeastern part of the country (Longworth and Williamson, 1993), the policies of national and local governments are largely to blame for the rangeland degradation that eventually led to irreversible degradation of the whole ecosystem, or desertification.

Rangeland in the PRC is State owned. Since the reforms in 1978, pastoralists have, in principle, occupied rangeland according to a system of contracts and leases. Very often the location of the "leased" land is not specified, meaning that in practice rangeland is treated as common grazing land. Incentives to invest in and improve pastures are minimal. When a piece of land is specified in a lease, the land's use and ownership may still be re-allocated or re-assigned by the Government. Such arbitrariness also discourages sustainable management or private investment and encourages shortsighted exploitative behavior on the part of the lessee. The responsibility for setting the terms and conditions of pastoral leases has been delegated to local governments, with the result that these terms and conditions vary widely from place to place. For example, leases are granted for as little as 5 or as many as 15 years by different local governments. Again, shorter lease periods decrease the incentive for investment in the land and the adoption of sustainable practices. In India, similar problems have occurred. Nationalization of arid rangeland has converted a community-managed "commons" into an open-access system (Steinfeld, de Haan, and Blackburn, 1998). As a result, the common resources were degraded by 30 to 50 percent over a 30-year period, and the number of grazing days was reduced.

The same land tenure problems occur in Central Asia. Crucial to the problem of rangeland degradation are public policies that have led to various uncertainties, and incentives that have induced operators to behave in an exploitative manner. This has, in turn, placed a constraint on finding possible technical solutions that might help to improve the productivity and sustainability of the pastoral system.

A holistic approach to rangeland management would include an evaluation of the influence of public policy, the role of traditional/communal rangeland management, and the potential technical solutions that may be employed. A policy process, at the county or provincial level, that involves participation of pastoralists and communities, as well as governments and technical professionals, would provide a framework from which the management of pastoral systems could be improved.

As far as legal recognition of individual rights to decision making is concerned, Myanmar is a notable exception. Although various legal forms of land tenure including ancestral holdings *(bo bwa bain)* exist, in practice the Government can reallocate land without compensation and designate it for rice production, limiting farmers' ability to make crop choices. In 1994/95, designated paddy land accounted for about 54 percent of the area sown to rice (US Embassy, Yangon, 1996). The Government has also diverted land for bean and pulse production. The areas for this have increased in response to liberalized trade. This system has reduced incentives to increase land productivity. In spite of the abundance of fallow and uncultivated land and of water, it has been estimated that there are 12 million landless laborers in Myanmar (ibid.).

Land tenure systems also affect conservation behavior. It was discovered that the land tenure system in the Philippines of three-year cash leases discouraged soil conservation practices. In the upland areas where land tenure security is absent, farmers lack incentives for the adoption of erosion control practices. Incentives in the form of input subsidies and marketing infrastructure without land security tend to further aggravate land degradation. The types of rights that need to be given are those that harmonize the goals of increased conservation efforts and productivity but minimize area expansion. Increasingly, governments are turning land outside protected and fragile areas over to local inhabitants. The issue now is how to deal with settlements in headwaters and fragile ecosystems. Even in land-abundant countries such as Thailand, the approach of removing settlements from fragile ecosystems is met with strong resistance. This is compounded by the difficulty of locating new sites for resettlement. Other policies, such as nonfarm employment and education for the younger generation in order to create more employment options for them, are needed to complement land policy.

In the more vibrant economies in Asia, where land speculation has been prevalent, especially before the recent financial and economic crisis, the failure to curb windfall gains either through capital gains tax or taxes on unused land encouraged land sales and forest encroachment. Appropriate tax instruments are needed to reduce pressure for opening land frontiers while unused land is still available.

Forest Policies

The state of natural forests in most Asian countries can best be described as critical. Natural forests in the resource-rich countries in Southeast Asia are dwindling rapidly. The PRC and India have already lost most of their natural forests. Much of this poor performance is a result of outdated policies and institutions. Forestry policies in many Asian countries are legacies from colonial administrations of centralized and bureaucratic regimes and are oriented towards production. In many cases they are public land policy rather than forest conservation policy.

Forestry agencies in Asian countries specialize mainly in extraction and production. With changes in the socioeconomic and political environment as a result of population pressure on forests and an increased awareness of ecology and biodiversity, these agencies found themselves without adequate personnel and expertise to cope with people living in the forests and with issues concerning protected forest areas. In some countries, the forestry sector is controlled by some specific interest groups. This leads to a lack of willingness to adopt new responsibilities for conservation. Protection is geared towards protecting State revenues and not sustainable forest functions or the livelihood of those depending on forests. State monopolies over forests have deprived local communities of access to resources for their livelihood.

Sectoral policies that encourage deforestation and overlogging include the underpricing of concessions, promotion of large-scale plantations in tropical rain forests, and forest policies that deprive the surrounding communities of de facto usufruct rights to forest resources and bestow monopoly rights on State forest management. In Cambodia, low royalty rates understate the true economic value of natural resources, leading to revenue losses of $100 million per year (Royal Government of Cambodia, 1997). In India, legal restrictions on the harvest and sale of tree products have reduced the incentive to grow and care for trees. In addition, the policy is deemed to affect soil fertility because manure has to be diverted from farm use to fuel use as fuelwood becomes scarce (Chambers, Saxena, and Shar, 1989).

Although forests in Asia are protected by law, they are de facto under open access owing to a lack of resources and personnel, except where local institutions for forest management exist (Box III. 3). Moreover, a broad range of nonforest sector policies can encourage new social and environmental objectives leading to deforestation, for example policies that support an over-expansion of agriculture either through trade protection or input subsidies.

Box III.3 Forests for the Grass Roots

In West Bengal, 3,000 communities are protecting forests that have a total area of more than 3 million ha. Forest legislation (1989) in West Bengal was amended in 1990 to empower local committees to manage forests. Forest Protection Committees (FPCs) were given the usufruct rights to royalties for fruits, flowers, grass, leaves, and one quarter of the timber (aged over 10 years) produced after 5 years of protection.

A case study of Bhagabatichak in West Midnapur has shown how a small community can design institutions that make both forest and communities more viable. Families in Bhagabatichak are mainly landless farmers and recognize that further degradation will affect their future livelihood. Common property rules are specified and monitored. Occasional raids against illegal activities are also undertaken. Village rules specify harvest schedules for six species of fuelwood. Forest floors are disturbed only in the dry season.

By 1995, the FPCs had replaced official authorities in forest protection. The West Midnapur Forest Division conducted an experiment to compare forests under community protection with open-access forests. It was not surprising to find that the extraction rate was lower in the community-managed forest as the people had gained a more secure supply of fuelwood.

Source: Thapan (1998, p. 262-3).

Protected areas in Asia are not well selected and are poorly demarcated and managed. Conservation staff are inadequately trained and biodiversity knowledge specific to individual parks is scarce. Financing options other than budgetary sources for protected area systems are rarely explored. As a result, protected areas have unduly become a budgetary burden.

The conventional policymaking process has an inherent bias against forest protection. Cost-benefit analyses of development projects that encroach into forests tend to underestimate the value of forests, especially their ecological functions and the future value of their biodiversity (Mingsarn, 1995). Empirical studies suggest that the nontimber value of forests is substantial (Reid et al., 1993; ADB, 1995a; Bann, 1998) and that there is a general willingness to pay to protect forests (Mingsarn et al., 1995). Other studies have indicated that benefits from clear-cutting in natural forests are only one seventh of those from sustainable yields of timber and nontimber products (McNeely, 1998 p.5).

This lack of proper valuation and neglect of the ecological and biodiversity functions of forests leaves their protection as a low priority in government development agendas, which in turn has biased government decisions for public investment for the protection of forests. In Thailand, forest land is considered reserve land for development, tourism, military bases, dam construction, and even garbage disposal. The cost-benefit analysis of a dam construction project in protected areas, for example, often excludes the cost of ecological and biodiversity loss, while timber extracted from the protected area is considered a benefit.

Finally, despite their high biodiversity, developing countries in Asia lack funds, personnel, and knowledge for protecting forest resources, for prospecting, and for establishing a systematic information and knowledge database for future use. A cost-benefit sharing system on a global basis would be needed in order to maintain the global commons (Mingsarn, 1997). Moreover, in the forestry agencies of most developing countries, especially those where extraction activities remain active, R&D personnel tend to lack support and have low morale. In addition, scientific capacity related to forest production in Asia is deteriorating.

Water Policy and Institutions

The principal challenge for a water management strategy is to design institutions that are responsive to changing needs. As the need for water for agriculture increases, most governments concentrate their efforts on the provision of water. Large investments in water resource development have led to the establishment of correspondingly large bureaucracies and industries specializing in water provision. Therefore, water policy and institutions in Asia deal mainly with the provision of water for irrigation, and the agencies are staffed by engineers preoccupied with construction and supply management. It is not surprising to find that irrigation policy in many Asian countries tends to be biased in favor of large-scale infrastructure developments such as dams (Mingsarn and Ammar, 1997; Vyas and Reddy, 1998).

This top-down approach to water resource development has been increasingly countered by public protests organized by NGOs. The increase in social, environmental, and political costs related to dam construction has made large-scale dams a less viable option for water resource management, not only in economic and social terms but also politically. A study of water institutions in 11 countries, including the PRC, India, and Sri Lanka in Asia, indicated a shift of the key issues being addressed from water resource development to water resource allocation and water quality.

Water management in most Asian countries is fragmented and sectoralized. For example, in India, surface irrigation, underground irrigation, drinking water, water supplies, and pollution control are under independent agencies and coordination is often not very effective (Moench, 1994). Another example is Thailand, where water is partially controlled by the Royal Irrigation Department, but there are more than 20 other agencies that also oversee water, under different pieces of legislation. Thus, water conflicts between sectors and between upstream and downstream users are difficult to resolve.

At present, planning for and implementation of water resource development is on a project basis. Increased competition for water, complexities in water-use patterns, inter- and intrabasin water diversion possibilities, and rising conflicts have necessitated "bioregional" planning at the basin level, with wider participation by water users and stakeholders (Mingsarn et al., 1999). The basin development planning process used by the Mekong River Commission, incorporating basin-wide stakeholder participation, is a move in the right direction.

Water institutions and policies in the PRC are relatively advanced in comparison with those in other Asian countries. This has happened for three reasons. First, the PRC is both flood and drought prone. Mitigation of effects of flooding has been a major concern from ancient times. Second, water resources in the PRC are unevenly distributed, although in terms of per capita availability and the percentage of withdrawal, the country cannot be considered water scarce given that its overall water resources per capita are about 2,100 m^3 per year. While most of the irrigated areas and population of the PRC are located in the north, it is in the south where water is most plentiful. For example, in the Hai, Huai, and Huang River basins, where 34 percent of the PRC population and 42 percent of its irrigated land are located, the total per capita water resources are only 310 m^3 per year (Saleth and Dinar, 1999). Third, soil erosion and water pollution in the PRC are particularly severe.

Water policies and institutions in the PRC are geared to solving these problems. In 1988, the Water Law was passed and the Ministry of Water Resources and Power was established. Under the Water Law, water is the property of the people and a clear distinction is made between user rights for people and the allocation rights of the administration. The river basin has been recognized as the basic unit of water management. Water Conservancy Commissions were formed to manage intraprovincial river basins and lakes. In 1997, the

PRC further unified its water resource management policies by enacting the Law of Flood Control and declaring a National Policy on Water Pollution Control.

Throughout Asia, surface water is under an open-access regime. Water resources are free or underpriced, leading to wasteful use. In areas where agricultural intensification is made possible by pumping irrigation, such as in India, overpumping has lowered the water table, such that households with shallow wells are deprived of water. Again, those who suffer tend to be the poorer farmers. In this case, underground water is open access, but the supply and demand imbalance is such that appropriate management of this resource is now a necessity.

When water is abundant, an open-access regime is appropriate. It is also considered an equitable regime. However, as the competition for water intensifies, an open-access regime can no longer produce equitable results. Water-related laws often do not explicitly specify quantitative restrictions on individual or group withdrawals. Water is disproportionately extracted by those who have more money, technology, labor, and power. Those who have access to free water often engage in wasteful uses. In many countries, allocation principles and mechanisms are needed, especially in times of shortages. Water agencies, which are accustomed to simply providing water, have found it difficult to devise allocation methods that are efficient and acceptable to all.

The multiple-use conflicts indicated in Chapter II imply the need for a comprehensive water administration body that can coordinate the needs of different users and establish acceptable and effective allocation principles, provide dispute settlement procedures, and maintain quality control standards while leaving the day-to-day management and dispute settlement to local institutions.

In a number of Asian countries, national guidelines on water use provide directives for pricing that reflect the scarcity value of water. In practice, except in Japan, irrigation water fees are not high enough to influence crop choice or irrigation practices at the field level. In India, surface water irrigation rates only cover operation and maintenance costs. Moreover, electricity charges for pumping irrigation are subsidized. In Pakistan, the establishment costs of irrigation are recovered by the sale of land in the irrigated areas and, as in India, water charges are meant to cover only operation and maintenance costs. In practice, water charges are too low even to maintain the system.

In Thailand, water prices for canal irrigation and underground water bear no relation to their actual cost. In the PRC, pricing has gained an important role in allocating water between and within sectors and also as a means for penalizing polluters (Chen, 1992). A permit system for withdrawing water is expected to be in use by 2010. Reforms are in progress that recognize variations in supply conditions in different regions.

Pricing alone does not guarantee efficient use if prices are tied to the cropping area or even type of crop. Farmers who pay flat rates for water tend to use or hold as much water as possible. In this case, proper pricing and education on water problems related to excessive irrigation will have to be implemented jointly to assure proper use of water resources.

Private water markets are the main feature of Indian tube-well irrigation. Elsewhere, markets for irrigation water have not been developed. In Thailand, where competition for water use is increasing, especially in the northern basins, the development of water markets is not easy. This is due to very small holdings; a mixed tree and annual cropping system, which implies different levels of water commitment; the similarity of crops between deficit and surplus areas; and potential political resistance from landowners (Mingsarn and Ammar, 1997).

In most Asian countries, pricing mechanisms need to be reformed, although water pricing is not a panacea. Rehabilitation, better maintenance, and management of the existing irrigation systems are necessary. Improving irrigation effectiveness is imperative. In many cases, large-scale water resource development projects are not a priority. In Cambodia, the existing "colmatage" systems, which consist of canals dug through the natural levies of the rivers to allow floodwater to inundate and fertilize rice fields naturally with nutrient-rich silt, need to be restored (Benge, 1991). In Viet Nam, the priority in water resource development is rehabilitation of the existing system, and improved drainage and water control against salinity intrusion, rather than new large-scale development. A simulation study by the World Bank suggests that irrigation investment in Viet Nam would have an insignificant effect on poverty and the smallest impact would be in the Mekong Delta. This is due to the peculiar nature of the Delta, which suffers long periods of inundation and seasonal shortages, and contains low-quality acid sulfate soils (Van de Walle and Monhindra, 1995, cited in Litvak, 1995).

Irrigation authorities will need to broaden their perspective and consider other management options and instruments needed to supplement the current command-and-control regimes and supply orientation. An incentive or penalty system needs to be established to encourage the use of water-saving technologies such as drip irrigation. In addition, stakeholder participation and involvement in water resource management could increase both efficiency and equity and could also reduce the budgetary costs of operation and maintenance.

In the mountain areas of many Asian countries, where topography and hydrology permit, irrigation systems have been developed as a common property of communities, and management is generally recognized as being efficient. In Thailand, group- or community-managed irrigation networks have been integrated into public irrigation systems. Increasingly, water-managing communities are being considered as important components in water resource management and need to be recognized, institutionalized, and given greater responsibilities. In India, villager participation helped to rehabilitate and maintain communal water resources (Box III.4). In relation to large-scale water resource planning and management, institutional mechanisms and capacities have yet to be developed for the assessment of social and environmental impact. These capacities need to be developed not only by environmental agencies but also by water agencies and the private consulting sector.

Box III.4 Villager Participation in Water Resource Rehabilitation in Ralegan Siddhi, India

Situated in the State of Maharashtra near the city of Pune, the village of Ralegan Siddhi has undergone an amazing transformation in less than 20 years (AVARD, 1993). By 1976, Ralegan Siddhi's agrarian economy was being ravaged by massive soil erosion, deforestation, recurrent droughts, and land degradation. There was also an acute water shortage. The water table was low, and all water (including that for what irrigation there was) came from wells that went dry in the summer, such that even drinking water had to be brought in. Under these conditions, only 30 percent of the village's food grain requirements could be met. The social consequences were equally devastating. Able-bodied men had little choice but to leave the village to look for work, while locally, illegal liquor establishments were set up. By 1976, Ralegan Siddhi had 40 liquor establishments, for a population of only about 1,200 people. Some 85 percent of the local population became addicted, with the problem even reaching schoolchildren.

The reversal of Ralegan Siddhi's fortunes began with the dedication and work of one individual, Anna Hajare, who had retired from the army in 1976 and returned to help make a difference in his home village. The most pressing issue was the acute water shortage. Water conservation measures were instituted, such as bunding, land shaping, land grading, and the building of water tanks and small check dams for storing rain water. Afforestation and pasture development helped to regenerate the local vegetative cover and the village ecosystem, controlling erosion, minimizing runoff, and permitting the development of previously unusable land. Bore wells were dug for drinking water, eliminating the water-borne diseases common previously. The new dams and water storage facilities helped to raise the water table, allowing new wells to be dug for irrigation. A lift irrigation project was completed in 1986 with the result that by 1993, 447 ha of land could be irrigated,

(continued next page)

Box III.4 (continued)

compared with only 25 ha in 1976. Although water is shared communally, there are various ways by which heavy users compensate those who use less, and this de facto user-pays system serves to enhance responsible water use.

These achievements required the voluntary efforts of the entire community, and the worthiness of their projects enabled government financial assistance on many occasions. As all helped, all have shared in the benefits and now Ralegan Siddhi enjoys not just abundant water and vastly increased crop yields (Ralegan Siddhi is not only self-sufficient but now supplies neighboring villages), but all the liquor establishments have closed down and the social ills are now mostly a memory. The local economy has improved to the point where the men who had previously left have returned. Through its remarkable achievements, Ralegan Siddhi stands out as an example of how the residents of a community can improve their lives through their own actions, and that it can all start at home with the efforts of just one community member.

Source: AVARD (1993).

Coastal and Ocean Resources Policy

Approaches to fisheries development in most Asian countries to date have largely been extraction oriented rather than conservation or sustainability oriented. Thus, as elsewhere, fishery policies have traditionally stressed increasing production goals, especially in marine fisheries, while the conservation and socioeconomic aspects of fisheries have received minor attention. In recent years, there has been "a shift toward conservation and ecosystem based management from traditional exploitation and stock or species based management" (Ahmed, Delgardo, and Svedrup-Jensen, 1997a). Yet for most countries the dilemma between increasing production to meet growing demand and practicing self-restraint and conservation remains a difficult one to resolve. In the absence of adequate information about the status of fishery resources, policies are in most cases ad hoc in nature (Ahmed, Delgardo, and Svedrup-Jensen, 1997a).

Yet resource rehabilitation has been an ongoing process, and many countries have been placing increasing emphasis on conservation, although the efforts are often too little or too late. Given the weak institutions and lax law enforcement in many Asian countries, success has been mixed. Fishers frequently ignore regulations on mesh sizes or gear use, and enforcement is often costly and difficult. A few successful examples have been reported, however. For instance, examples from the Philippines and Cyprus show that seasonal fishing bans can bring about sustainable production increases of 100 percent in less than 18 months (Garcia, 1986, cited in FAO, 1997b).

Often such successful efforts lack consistency. In Thailand, for example, a similar seasonal fishing ban over an area of 26,000 km^2 in 1984 resulted in increased catch rates in just a few years. However, the ban was later lifted for beam trawlers and anchovy purse seiners, purportedly to ease the plight of the fishers using those methods. The ban was really lifted because of a growing export demand for shrimp, which are caught by beam trawlers. Shrimp fishing leads to more bycatch and waste; thus, demersal catches will once again decline rapidly (Boonlert, 1994).

This difficulty is compounded by the fact that the local stakeholders who benefit from the extractive approach to resource management tend to have connections with politicians or administrators in control of policy decisions (Mingsarn and Pednekar, 1998). Therefore, attempts to enforce no-fishing zones or conservation measures stipulated by law are sometimes neutralized by executive decisions. In many Asian countries, political reform has become a prerequisite for economic reform.

Another major issue in aquatic resource management is the management of multiple-use conflicts. In coastal areas, such conflicts pose the greatest challenge. As mentioned earlier, conflicts are increasing between small-scale and commercial fishers, between marine aquaculture and trawling, between shrimp farming and rice farming, between tourism operations and small-scale fishers, etc. These conflicts are not limited to private entities, but occur between public conservation agencies and public production-oriented agencies. This has led to lax law enforcement and situations where outcomes are determined by mob protests and political power plays. This renders the rehabilitation and management of resources difficult.

A major reason for multiple-use conflicts, overfishing, and the resultant stock depletions, is the continuing de facto open-access nature of most of Asia's fisheries, despite growing attention to the worsening situation at both international and national levels in the form of international agreements, conventions, and law reforms. The few notable exceptions include the Japanese system of fishing rights in coastal zones and licenses for offshore fishing (Yamamoto, 1998).

More recently, however, governments in a number of countries have tried to develop various measures to restrict access by giving user rights to local fisher communities or by designating fishing zones. Among these are the ongoing efforts in Bangladesh towards the management of inland fisheries through the involvement of government, NGOs, and local communities (Ahmed, Capistrano, and Hossain, 1997b). Attempts are being made in various countries to define boundaries for coastal fishers. These range from 3 km from the coast in Thailand for fishers with boats smaller than 10 GRT (Ruangrai, 1997, cited in Mingsarn and Pednekar, 1998), to the 15-km zone of the Philippines' municipal fisheries sector for small-scale fishers with boats of 3 GRT or less (Deb, 1997). However, despite these measures, boundary violations and the resultant disputes between small-scale and commercial fishers are common occurrences in most countries, largely due to the lack of proper enforcement and monitoring systems. Moreover, license and registration fees for fishing boats and gear generally bear little relation to their extractive capacities.

Along the lower Mekong River in Lao PDR and northern Cambodia, local fishing rights have been a customary practice. In Tonle Sap, Cambodia, a fishing lot system has been employed for over a century (van Zalinge et al., 1998). Under this system, the fisheries are monitored to protect against theft and to protect wildlife habitats, flood plains, and mangroves; extraction tends to be consistent with maintaining long-term ecological balance. However, the recent influx of migrants due to improved access to the area has created an open-access system and created conflicts between many operators. Institutional reform that takes into account the existing social structure has become an emerging necessity.

Rehabilitation is also important for inland fisheries. Unlike marine fisheries, where rehabilitation usually focuses on reducing fishing effort, the focus in inland fisheries needs to be placed on reducing the pollution of degraded water resources. A number of countries have also launched programs for seeding inland water resources with hatchery-bred juvenile fish, etc. in order to improve production (Coates, 1995).

Owing to the many stakeholders involved, coastal and aquatic resource management needs to be participatory in nature. Fishery management after all is "a balancing act between the requirements for biologically sustainable resource use, economically optimal exploitation patterns and their social acceptability by the involved parties" (van Zalinge et al., 1998, p.9).

IV CHALLENGES AND OPPORTUNITIES FOR ENHANCING AGRICULTURAL GROWTH AND SUSTAINABILITY

I n order to achieve sustained increase in agricultural productivity, a number of major challenges need to be addressed. They have been tackled in the past with some success, but the complexity of the issues involved requires that a much more coordinated effort be made than has previously been the case.

CHALLENGES

Population Growth , Poverty, and Environmental Degradation

The rates of annual population growth in Asia in the last two decades have been slightly higher than the world average, with South Asia showing the highest growth (Annex Table A1). Population growth tends to be high among the low-income and low-education groups and among those residing in remote areas, owing to the lack of access to family planning services. Although even the more populous Asian countries have been able to meet increased demands for food despite growing populations, the region is not free from hunger and malnutrition.

In poverty stricken areas, population increases tend to accelerate the short-term extraction of natural resources for sustenance. At low levels of income, natural resources are treated as consumer goods. Population increases also produce a large pool of labor that has a low opportunity cost. Together, these can trigger a vicious circle of poverty and environmental degradation. In countries where growth is stagnant or slow, these problems tend to become even more severe.

Poverty may have an impact on the environment in two ways. First, poverty influences the timing of consumption decisions, causing them to be biased to the short term. Second, poverty alleviation may have positive or negative impact on the environment, either directly or indirectly. In the first instance, poverty prevents people from taking a long-term view and from investing in benefits that are not immediately realized. In other words, poverty increases the discount rate and renders short-term gains more attractive. For example, poor farmers may overextract trees and convert them into charcoal for immediate cash and consumption needs, despite the fact that the same trees would yield a greater range of benefits over time. In the highlands, poor farmers can little afford to spend time in water and soil conservation because they have to meet their day-to-day need for food. A study of a labor-surplus economy in a semi-arid area of southern India indicated that farmers viewed water and soil conservation as an expensive and labor-intensive undertaking (Kerr and Sanghi, 1991, cited in Gill, 1995). The situation is often exacerbated by the lack of security in land use.

Population growth often increases poverty and causes migration in the search for new farm land. In a country where arable land is scarce, poor farmers are often associated with environmental degradation. A study in the PRC estimated that about 85 percent of the rural poor, or about 85 million people, reside in degraded areas (Yin Runsheng, 1997). Continued population growth places further pressure on the existing capacity of the land, forcing farmers to cultivate their crops in higher and steeper areas (Box IV.1). Often the poor reside in areas unfit for agricultural production. The challenge here is to

Box IV.1 The Traditional Knowledge of the Ifugaos

Traditional knowledge had led to sustainable agriculture in the past. Shifting cultivation, for example, permits the regeneration of the soil during the fallow period. The wet rice cultivation system is a closed system whereby the rice stalks and the manure from draft animals replenish the soil (Rattan, 1994). The amount of nutrients removed from the soil is negligible. However, this knowledge was developed at a time when the population was small and increased only slowly. Today, traditional knowledge is still applied but has been modified to meet increasing needs.

The Ifugaos of central Northern Luzon, Philippines, routinely manage complex cropping systems comprised of terraced rice, swidden, private woodlots, and communal grasslands. The cropping of rice is combined with fish culture, and with maize and tubers on the bunds. Swiddens are used to produce supplementary food crops such as maize, pulses, and tubers, and are left fallow for 7 to 8 years. Secondary forests on steeper slopes, where cash crops such as coffee and rattan can be grown among the *Dipterocarpus* species, are divided into plots for each family. A more distant forest is set aside for communal use; the use of resources is governed by community rules.

The Ifugaos have been able to augment their food supply to meet increasing needs while maintaining the long-term sustainability of their system. Annual soil loss from the woodlots and rice terraces is estimated at 0.2 t/ha, a rate that is considered very low. Erosion from swiddens is estimated at 10 t/ha per year.

Sources: Toribio and Orno (1995); Ticsay-Ruscoe (1995); Magliano and Librero (1998).

find sustainable agricultural production systems and nonagricultural strategies that would generate gainful employment and income for the rural poor.

It is widely recognized that poverty alleviation is imperative if sustainability is to be achieved. A number of policies and associated public expenditures have been directed towards poverty alleviation, with varying levels of success. Some poverty alleviation policies may have an impact on the environment or create an incentive system that works against the environment. A recent study in India (Fan, Hazell, and Thorat, 1998) showed that government spending on roads has had the biggest impact on rural poverty reduction, and that the impact of road construction is almost twice that of government spending on agricultural R&D. Education, rural development, and irrigation (in that order) also have positive but lesser impact on rural poverty reduction. However, spending on fertilizers and other subsidies was not taken into account in the study.

That roads are an important part of infrastructure and have the greatest impact on poverty reduction is not contested here. However, the environmental implications of expanding road networks should be noted. Poverty-stricken populations are sometimes located near biodiversity-rich areas and tend to depend on this common pool of resources for their livelihood. Lack of access is a common characteristic of both poverty and biodiversity. As access improves, the poverty situation tends to lessen because of better access to education, health services, and income-generating opportunities. Biodiversity, however, tends to deteriorate owing to the commercial-scale activities that are made possible with improved access to such areas. Precautionary measures to protect biodiversity and to support local institutions that conserve local biodiversity are necessary if road infrastructure policies are to become truly win-win solutions. Granting use and protection rights to local communities is one management option that could improve the situation for both communities and biodiversity conservation.

Often agricultural policy is used to alleviate poverty in rural areas. In the name of helping the poor, subsidies on inputs such as credit, seeds, chemicals, electricity, fuels, and water are offered indiscriminately to all farmers. This results in wasteful use of resources, increased budget burdens, and, in many instances, increased debts for farmers. All of this leads to environmental degradation, e.g. from the overuse of chemicals, the overpumping of groundwater, etc. In addition, governments may guarantee output prices and protect crops against imported substitutes. Subsidies and price distortions resulting from the aforementioned policies also provide confusing signals concerning technical change, and hinder the adoption of technologies that are more resource saving and environmentally more benign. Under the guise of poverty reduction, governments have revoked prohibitions on trawling and pushnetting, allowed fishing in protected areas, and given marginal lands in fragile ecosystems to the landless. It is not surprising that in the end, the poor are often seen as the culprits of unsustainable production and environmental degradation.

It is argued here that the above policies, which are, in fact, lose-lose policies, arise because poverty is often defined too narrowly as an inadequacy of material wealth. While this is true, it does not completely encompass the concept of poverty, which ought to include social and political characteristics such as the lack of access and rights to land and common resources, vulnerability, social insecurity, dependency, and a lack of choice. Only when the definition of poverty is viewed in its broader sense can one understand why a policy of government handouts often fails both to alleviate poverty and to promote sustainable production. The poor need not simply credit, material inputs, and one technology that fits all; they also need rights, access, the ability and opportunity to make choices, knowledge and understanding of local conditions, and technology fit for local use. These needs are extremely difficult to meet under the existing centralized, top-down bureaucracy that typically characterizes so many agricultural agencies in Asia.

Another challenge related to poverty and agricultural sustainability is that although the capacity now exists in Asia to produce more than enough food to feed its population, increased production will only occur when there are reasonable profit margins. As the cost of production rises because of greater competition for resources and environmental degradation, the price of food, especially rice, may also rise. Should food prices be allowed to increase in order to provide an impetus for increased food production, the means must be found to protect the nutritional standards of the poor, especially those employed outside the agricultural sector.

Less Favorable Environments and Fragile Ecosystems

Population growth in Asia has placed and will continue to place immense pressure on land, leading to agricultural intensification in less suitable areas, and to the opening up of new lands and the clearing of forests for cultivation. This means encroaching on natural forests, intensifying agriculture in less favorable environments (LFEs) and fragile ecosystems, or shortening fallow periods in shifting cultivation.

Less favorable environments are characterized by less than optimal growing conditions, e.g. areas where precipitation is low and unreliable, where the growing season is short, where soils are poor, and where topsoil is depleted. LFEs can occur naturally or result from the mismanagement of fragile ecosystems. The term "fragile ecosystem" is also often used to define areas or characteristics of sites that are "too dry, too steep, or lacking in nutrients" (WCED, 1987). The more recent literature tends to stress the relationships between society and nature rather than biophysical characteristics–a dynamic rather than a static interaction–and the mismatch between human uses and system capacity (Turner II and Benjamin, 1994). Fragile land has been defined in terms of two properties: environmental sensitivity and resilience. Truly fragile ecosystems are those that are "highly susceptible to biophysical deterioration and do not really recover" (Turner II and Benjamin, 1994, p.113). For the purpose of this study, those ecosystems where productivity deteriorates rapidly and that are costly to restore are called fragile ecosystems. Degradation also causes negative externalities that cannot be compensated for by gains from the change in land use.

Asia 's uplands are covered by relatively less fertile acidic soils that are highly erosive. In Southeast Asia alone, approximately 188 million ha or 39 percent of the total land area is acidic upland. In Lao PDR, this proportion reaches 66 percent (Garrity and Agustin, 1995). Population pressure on these uplands has considerably shortened fallow periods and accelerated environmental degradation.

In the PRC, fragile lands are common in four areas, the Loess Plateau in Shanxi, Shaanxi, and Gansu provinces, the red soils areas, the northeastern plain, and the northwestern grasslands. Together, they account for 70 percent of the PRC's land area (Rozelle, Huang, and Zhang, 1997). These areas are partly but not entirely fragile and degraded. Some parts have been cultivated continuously since ancient times.

Many factors contribute to encroachment on or the inappropriate use of fragile lands, ranging from national policies (e.g. on migration, Box IV.2) to sectoral policies, such as the replacement of narcotic crops. Other factors include commercialization, natural population increases, land speculation, and attempts to increase the productivity of fragile land.

The isolation of some of the highlands in the montane regions of Southeast Asia, for example, provides natural protection for illegal crops, especially opium. Although opium is a traditional crop that is used as a medicine among the hill tribes, the commercialization of opium in the "golden triangle" was driven by the fact that opium is also a narcotic and is banned in most countries. The replacement of opium by other, higher-value crops, mainly through horticulture, creates second-generation problems. The agricultural intensification that horticultural crops require is made possible by huge subsidies from governments and international agencies. Some of the crops have spread beyond project areas (where soil and water conservation are part of the technology package), to the hills and mountains where there is inadequate soil and water conservation (Kanok et al., 1989, 1994).

The degree of soil erosion from steep farmland depends on the type of crop being grown, the agricultural practices used,

---

Box IV.2 Exodus into Fragile Land

Most of the development literature has emphasized rural to urban migration, but in the case of fragile land the reverse is often true. Population pressures in fragile lands often stem from a migration of the mainstream population from relatively dense population centers into indigenous and sparsely populated communities. For example, in the PRC's Xinjiang Uygur Autonomous Region, the exodus of mainstream Han Chinese increased the Han population from 291,000 in 1949 to 2.1 million in 1962, and finally to over 5 million by 1990. In the Ardos Plateau of inner Mongolia, the Han population contributed to a 225-percent increase in population over two decades in response to the Government's policy of regional development and land-use intensification.

In Viet Nam, 5 million Kinh, the main ethnic group, have moved from the lowlands to the central areas of the highlands during the last 30 years, and the rate of migration has accelerated since 1980 (Sam, 1994). Migration into the highlands of Chiang Mai, Thailand, is estimated at 12 percent per year (Kanok et al., 1994), leading to intense competition for water resources between highland and lowland farmers.

---

amount of precipitation, and topography. Annual losses range from less than 2 t/ha if conservation is practiced, as in the case of northern Thailand, to about 100 t/ha for grazing land in Nepal and in the uplands of the Philippines (Shah, 1997).

Inappropriate agricultural intensification can do great harm to LFEs. For example, the conversion of low-productivity grazing land to higher-productivity cropland is sometimes a self-defeating course of action. Grazing lands generally have few soil nutrients and are in areas of low and unreliable rainfall. The loss or decline in land productivity arising from the conversion of grazing land to cropland sometimes occurs because such conversions decrease the amount of land available for grazing, which results in overgrazing in the remaining rangeland. Moreover, the best grazing land with relatively high carrying capacity tends to be the first to be converted into cropland. The conversion is not always successful either, because of poor planning and execution, e.g. irrigated areas tend to be prone to salinization.

In addition, the damage from failed conversions is often difficult to correct. The Ardos Plateau represents one such failure. For centuries, this Plateau, a semi-arid to arid windy plain in inner Mongolia, supported a grazing economy. During the Great Leap Forward and the Cultural Revolution, attempts were made to intensify agriculture without any consideration being given to land conservation. This led to desertification in much of the region, and allowed dune mobilization to double between 1957 and 1977 (Huang et al., in press, cited in Turner II and Benjamin, 1994, p.125). Other disastrous consequences included catastrophic flooding and the transportation of sand from the Plateau, which accounts for 10 percent of the sediments in the Huang He (Yellow) River. Since the late 1970s, degraded cropland in the Ardos Plateau has reverted to pasture and forest or is under rehabilitation.

The issue here is that fragile ecosystems must be treated as a special category in agricultural development. Normal or conventional agricultural technology is not suitable for fragile ecosystems; neither are they suited to agricultural intensification. Yet economically viable technology, which must be particularly concerned with the environment, must be developed for fragile ecosystems to benefit those with no alternative employment residing in such areas. Moreover, a package of policies will have to be introduced to control population growth, to increase education opportunities and therefore increase future employment options, to generate nonfarm employment, and to encourage more environmentally friendly agriculture. Efforts to improve technology for fragile ecosystems have been made by ICRAF and some national governments, but more has to be done to keep pace with the degradation that is currently undermining system sustainability. In fact, agricultural policies and instruments alone will never be adequate for the achievement of agricultural sustainability. Community efforts may be needed for this purpose also.

The green-revolution technology package that is now a common management practice in the resource-rich or favorable environments is not suitable for fragile ecosystems. Neither is it applicable in LFEs. New technologies will have to be designed to help LFEs enhance agricultural productivity in a sustainable manner. There have been a few success stories in LFEs, however, demonstrating that even in these areas productive and sustainable crop management is possible. These successes also show that there is no one single solution that is widely applicable to LFEs, given the wide variety of problems afflicting them. This is in marked contrast to the favorable areas, whose similarities enabled the widespread application of the improved seed, fertilizer, and irrigation package of the green revolution.

To improve crop production in LFEs, institutional and social solutions often must accompany or precede technical solutions if the benefits from the latter are to be realized. An RD&E system capable of unlocking local potential and responding to specific local constraints is absolutely essential. Public investment in RD&E for LFEs is unlikely to generate the same rate of direct returns from crop production as that realized from RD&E for favorable environments. However, this investment could have other, more indirect, benefits, such as poverty alleviation; prevention of resource-base degradation due to salinization, erosion or desertification; and the slowing of deforestation.

Environment and Trade Issues

Debates on the relationship between trade and the environment have continued for several decades. Environmental groups tend to regard trade as a catalyst of environmental degradation. Insatiable demand of the affluent North, combined with the profit maximization objectives of traders in the South, has led to the unhindered depletion of natural resources and to environmental degradation.

Proponents of trade have argued that blaming trade for environmental problems is inappropriate. Apart from transportation activities related to trade, trade does not itself create environmental problems. Environmental problems occur mostly during production and consumption, and appropriate policy interventions need to be designed to tackle problems at the source. When carefully analyzed, however, domestic trade and environmental policies are often found to bear the responsibility for environmental problems (Box IV.3).

Contrasting with the fears of environmentalists that trade liberalization would spur more agricultural expansion into tropical forests, a simulation model of complete trade liberalization and instantaneous adjustments showed that the impact from production relocation on total food output would be quite small, amounting to a 3–8 percent increase in developing countries and 5–6 percent decline in output in developed countries (Anderson, 1998). A substantial production increase would come from North America and Australasia. In North America, where land has already been cleared, such trade policy reform would also shift livestock production from areas concentrated in industrial-type production to land-intensive systems that are less chemically intensive. The shift is likely to be from densely populated to more sparsely populated areas, resulting in a net decline in degradation. As long as proper environmental policies are implemented, countries can offset their marginal cost of degradation.

Continued attempts by those wanting to protect the environment are being made both at the international and national level to curb trade practices related to natural resource extraction, and to institute national and international governance regulating trade related to natural resources. At the international level, the General Agreement on Tariffs and Trade (GATT, now the World Trade Organization) rules are considered a major obstacle to efforts to strengthen the international governance of the environment. Despite the fact that trade is perceived by environmentalists as evil, the power of trade sanctions as an effective tool to force countries to behave in an environmentally friendly manner is well recognized. Even

Box IV.3 Thailand's Cassava Export:
Bad Policies not Bad Trade

Thailand's cassava trade is often used as an example of an industry that endangers the environment because cassavas are mostly grown in the recently deforested plateau of northeastern Thailand. As a tuber, the cassava is thought to deplete the soil of more nutrients than do most other cash crops grown in Thailand. In the 1970s and early 1980s, loopholes in the tariff system of the European Community (EC) rendered the combination of cassava products with soybean the cheapest livestock feed in the EC. As a result, the EC has become Thailand's major market for cassava products. The cassava trade is now perceived as having mined the country's resources for chicken feed. The feed-grain trade is also perceived as trading in environmental impact, involving a massive transfer of nutrients and often polluting the location of final use.

The deforestation in northeastern Thailand was the result of a mix of concessional logging, clearing of forestland to suppress communist insurgents, and a lack of enforcement of forestry laws. Once the forests were cleared, subsistence farmers moved in to eke out a living. Since the deforested land was considered public land, these farmers were denied ownership. Insecure ownership prevented them from making any long-term investment in the land, and encouraged extensive cultivation with a minimum of capital investment.

The cassava is regarded as a poor man's crop. It is hardy and can be grown even on poor soil. It is resistant to both drought and insects. Little investment in seeds or chemicals is required to grow cassava. This makes the cassava an appropriate choice of crop for the farmers in the northeast, who still remain among the poorest farmers in Thailand.

Thailand has a relative abundance of fertile land, but since cassava will grow even on poor land, the country's comparative advantage in the production and export of cassava pellets is the result of a good transportation network linking the fields

(continued next page)

Box IV.3 continued

to the factories and the factories to the shipping ports, not the result of mining soil fertility to produce cassava. Owing to the special circumstances created by the tariff system in the EC, Thailand enjoyed for some time a lucrative market, offering prices three times higher than elsewhere. It would not have been wise for Thailand to restrict its exports of cassava pellets, but the Thai Government could have done more by investing the excess profits from the cassava trade in replenishing the natural capital stock, in this case soils, or in other sustainable development.

Source: Ammar (1989).

environmentalists want to use trade sanctions as punishment for those countries lacking appropriate care and conservation measures. Trade proponents, however, are doubtful as to how trade sanctions, which are themselves welfare-reducing measures, will be able to lift the level of global welfare through environmental protection.

Over the years, numerous multilateral environmental agreements (MEAs) have been formulated to help protect and conserve the environment. A few of these MEAs have clauses restricting trade. Environmentalists are concerned that GATT Article 20, which allows members to use trade sanctions, is too narrowly interpreted to allow it to protect the environment, and it could neutralize the effectiveness of the MEAs. This issue is not discussed further here as it is only remotely related to rural Asia. However, attempts at the national level to protect the environment by strengthening domestic regulations related to trade have been prevented on the grounds that they would conflict with GATT rules. Actions preempted include high import levies on tropical wood in Austria, the revenues from which would go to a tropical forest conservation fund, and the

Netherlands Government initiative to limit lumber imports only to those countries that manage their forests sustainably. The most cited conflict is GATT's Panel of Judges' rule against a US ban on imports of tuna from Mexico on the grounds that the tuna were acquired by means (purse seine nets) that are dangerous to dolphins. More recently, the World trade Organization also ruled against a US ban on imports of shrimps from Thailand on the grounds that the shrimp harvesting methods employed could harm sea turtles. On both accounts, the US action was considered an attempt to impose standards on production processes. Under the agreement from the Uruguay Round, restrictions on standards for production methods are not allowed unless the production process in question affects product characteristics.

At times the US also uses trade restrictions against products to fulfill environmental objectives that are not directly related to the banned product itself. For example, in 1994 the United States banned imports in five product categories, including shoes and bags made from reptile skins; decorations made from coral, seashells, and animal bones; frog legs for human consumption; and goldfish and other tropical decorative fishes and feathers. This measure was a response to the use of tiger bones and rhinoceros horns in medicine in Taipei,China.

Phytosanitation control and technical harmonization are often imposed by the importing countries in order to protect consumers and to control pests and diseases. Agreement on the application of sanitary and phytosanitary measures in the Uruguay Round Final Act allows members to apply their own control on imports provided that the control is in accordance with international standards or is justified scientifically. The EC in particular is more stringent with phytosanitary controls and technical harmonization for agricultural commodities. For agricultural products, standards for residues are set in order to protect EC consumers. Certification is required for feed grains and other horticultural products to ensure that the commodities are disease and pest free. Veterinary controls are required for poultry. Moreover, exporters of fish products to the EC have to be from approved zones. These requirements pose important challenges to developing countries in Asia, which will have to upgrade their production processes to meet international standards. In the case of fisheries, these requirements will pose important constraints on rural poor in the coastal zones who earn their living from the preliminary processing of fishery products.

The above discussion indicates that for exporters of agricultural products, natural resource and environmental management as well as phytosanitation control will be increasingly important. Pre- and postharvest technologies for phytosanitation control will need to be developed and disseminated to small-scale operators including fishers. Governments will also have to pay more attention to conservation practices, especially in export-oriented sectors.

Recent international trade developments indicate a tendency towards lowering of tariff barriers, but increasing restrictions on environmental and health matters. The direction of change is much more predictable in EC countries and Japan than in the US. Exporters from developing countries will have little choice but to improve their domestic standards to meet market demand. For this purpose, improved facilities for landing and postharvest technology may be needed, for example, to assist small-scale fishers. At the same time, developing countries may want to investigate policy and institutional reforms that would help strengthen conservation and improve health conditions, and which could lead to a win-win solution, i.e. increasing export income while improving the local environment and health conditions in the importing countries.

OPPORTUNITIES

In discussing opportunities, two points need to be noted from the outset. First, in a highly competitive world, new opportunities are constantly being sought and pursued. New opportunities are often thought of as new ideas and new methods that will improve current situations. This way of thinking may be appropriate for the profit-maximizing private sector, in which inefficient firms or firms that repeat old mistakes are weeded out through competition. However, for the public sector the greatest opportunity is to learn from and to correct past mistakes. In attempting to achieve sustainable development, the opportunities are choices to not allow resource degradation to worsen, to avoid the wasteful use of public funds, to make policy reforms, and to invest in economically and socially high-yielding projects.

Second, improvements in agriculture tend to be automatically market driven. The public sector and international lending agencies are not able to identify commercially rewarding projects before the private sector does. Nor is this the role for the public sector. The role of the public sector is to set up fair rules for the game, seek opportunities for the small operator, and ensure that no one is left behind.

The focus here is on three issues. First, given the current resource and technology situation, what are the opportunities and sources for increasing food supply, or is there any scope for future growth at all? Second, will biotechnology and the so-called alternative agriculture be able to contribute to sustained agricultural growth? Third, who are the likely beneficiaries and losers from biotechnology?

Scope for Future Growth

Future agricultural growth in the current high-growth areas will depend on a sustained input of new plant varieties that will continue to break existing yield barriers. For rice, a new plant type is being developed by IRRI. It is expected to embody improved disease and pest resistance while increasing the yield potential by 25 percent and also improving grain quality. Another expected breakthrough is the development of apomixis (the capacity to set grain without sexual fertilization) in tropical hybrid rice, which will further reduce the cost of hybrid seed. Together, these would increase the yield potential by 50 percent. One current weakness in the new plant type is its inability to fill all of the grains fully, although it is expected that this will soon be overcome. The new plant variety is expected to be supplied to national breeding programs for field tests by 2000.

For wheat, despite continual and strong yield growth, several new innovations are in the offing. CIMMYT is working on a hybrid wheat variety that would break the yield barrier of existing varieties by means of improvements in chemical hybridizing agents, biotechnology, and a new plant type (Pingali and Rajaram, 1998). Elsewhere, wide crosses are being conducted between elite varieties and wild relatives to produce a "synthetic" wheat that can transfer desirable traits more easily.

For maize, highly favorable areas are likely to benefit from R&D by private enterprises. In the PRC, scientists are confident that the present 10 percent gain in yield potential of successive generations of hybrids will continue into the next decade.

In the shorter term, yield improvements can be realized by closing the yield gap that exists between experimental and actual yields. Even in the PRC, where actual yields are high, the gap between experimental yields and farmers' yields can be as high as 9 t/ha for rice and 11.5 t/ha for maize (Lin, 1998a). Closing this yield gap would require sustained investment in agricultural research, in the maintenance of soil fertility, irrigation systems, and strengthening the technological extension system. Appropriate incentives for farmers to adopt the new technology and maintain soil fertility levels would also need to be provided. Improving incentives for farmers in the PRC would also mean liberalizing grain prices such that the production of grain would again be a profitable enterprise.

Apart from further increasing the growth potential of the HYVs, opportunities exist for improving yields in LFEs. Although the early MVs were bred for use in favorable environments, later generations of MVs for LFEs have been bred by many national breeding programs. The yield potential of these MVs is more modest but they are hardier or better adapted to LFEs, e.g. deepwater or saline soils. Suitable varieties for flood-prone areas and acid sulfate soils in the lower deltas of Cambodia and Viet Nam will improve yields in these areas. Yield potential remains to be tapped in Viet Nam, Pakistan, and Lao PDR.

Other crops such as oilseeds, roots, fibers, and tobacco have also benefited from modern plant breeding techniques. The All India Coordinated Research Project released during 1982 to 1995 about 50 varieties of soybeans suited to specific locations. This has made possible an expansion of almost 5 million ha in soybean cultivation area, especially in the locations with the least irrigation. In Thailand, drought-affected areas have benefited from R&D on open-pollinated maize and cassava varieties by a local university.

The above opportunities are related to genetic improvement. Scope also exists for increased output in less intensive agriculture. More than half of Asia's riceland produces less than 3 t/ha. The potential exists for increasing this output through the use of better crop management, prudent water resource management, market and trade liberalization, and simply from an increase in price and profitability.

In Viet Nam, an additional 1.0 to 1.5 million ha could be brought into cultivation (20 percent of the current area). About 200,000 ha of land in the Mekong Delta could be converted from single cropping to double or even to triple cropping, if water drainage and salinity controls can be provided in order to reduce inundation in the wet season and water shortages from February to May (Mie Xie, 1995).

Enormous potential for an increase in food production exists in Myanmar where labor, uncultivated land, and water are abundant. This potential has so far been constrained by government policies. In Myanmar, agriculture remains centrally controlled and exports of rice are monopolized. The State provides directives for land use and designates areas for rice production, the timing of planting, and cropping intensities, but with little recognition of local conditions and potential. In addition, a quota below market prices for State purchases is specified for rice farming. Although rice outputs have increased under government direction, yields have been stagnant and export targets have not been met. Viet Nam, however, has been able to feed its large population while also becoming one of the world's major exporters. Its population is 1.6 times that of Myanmar while its arable land is only about 60 percent that of Myanmar.

At present, Thailand, which is currently the world's largest rice exporter, still produces an average rice yield below 3 t/ha. This is because Thailand has, from the very beginning, opted to improve the yields of local varieties that have a better eating quality rather than to adopt the HYVs directly. There is therefore enormous capacity for that country to increase its output, should the prices for or the margins from HYVs be high enough. In Indonesia, additional land for food production could be found on islands other than Java, including southern Irian Jaya.

Investment in R&D for the alleviation of onsite effects can be rewarding. The effective alleviation of salinization, largely by draining the soil with tube-well pumping and installation of drainage facilities, has allowed large tracts of salt-affected areas, especially in India and Pakistan, to reap the benefits of the green revolution. The Salinity Control and Reclamation Project in Pakistan has succeeded in increasing the area free of surface salinity from 49 to 74 percent, decreasing the area with severe waterlogging from 16 to 6 percent, and, by using pumped water for dry-season irrigation, increasing cropping intensity from 84 percent to 117 percent. The gross value of crop production increased by 94 percent in an area of 2.3 million ha (International Commission of Irrigation and Drainage, 1991).

There is also great potential for increased production of poultry and pork in peri-urban areas and aquaculture for high-income markets, but market forces have already provided the impetus in these areas. The role of the public sector in these areas should be to ensure that the external consequences of these industries are internalized and pollution is kept under control. The possibility also exists for combining bovines (cattle and buffaloes) with perennial tree crops such as coconut, oil palm, and rubber. A study in the 1970s indicated that if half the areas devoted to tree crops could be integrated with livestock, no new land would be required for an increase in livestock population of 25 percent (Payne, 1976, cited by Reynolds, 1995).

For fisheries, it is estimated that an increase of 20 million t in annual global fish production would be possible, if degraded resources were rehabilitated. FAO (1997c) indicated that an increase of about 16.1 million t may be possible in the Indian Ocean alone. Tuna fisheries in the Indian Ocean and the western central Pacific (South and Southeast Asia) also hold promise for higher production, particularly of skipjack and yellowfin tuna, but probably also of swordfish (Majkowski, in FAO, 1997b). There is also room for expansion in the western Indian Ocean, where newly recognized resources are available, such as the mesopelagics (e.g. lanternfish), whose global biomass is estimated at around one billion t (Pauly et al., 1998). In the western Indian Ocean, the estimates of lanternfish stocks vary from 1.7 to 20 million t, and after a long period of expectancy, commercial fishing for these species finally started there in 1996 (Shotton, in FAO, 1997b).

For inland fisheries, an increase in global production of no less than 5 million t in the next decade is considered possible. Freshwater aquaculture holds the greatest promise. The explosive growth of carp and tilapia farming implies that more growth potential can be tapped because genetically improved strains are emerging. Moreover, tilapia is also exportable, with markets already existing in the United States, Europe, and Japan. Currently, Taipei,China is the largest exporter, contributing around 156,000 t/year live weight (ICLARM, 1998). With more research and development effort, not only in pure science but also in processing and marketing, markets for freshwater aquaculture products can be expanded significantly. The expansion of aquaculture production from subsistence levels to commercial production for higher-income markets also requires the transfer of improved farm-level management and agribusiness practices. Small-niche and very-high-income markets also exist for low-input aquaculture, such as of high-value abalone and clams for east Asian markets.

Agro-based Industries

Most discussion on the sustainable development of agriculture has neglected the role of agro-based industries, despite the fact that their development could drastically affect long-term production and crop mix, as well as institutional arrangements with farmers. In addition, biotechnological innovations and new product developments could alter input requirements substantially. To date, there have been few studies on agro-based industries, except for activities that include primary processing and canning, which tend to generate relatively little added value in Asia. Secondary and higher-level processing tends to produce more added value. In Europe, the US, and Australia, primary and secondary processing can create products with high added value, for example wines, brandy, and distilled spirits. Such opportunities have not been sufficiently explored in Asia, although the region has a multitude of fruits, flowers, and herbs.

Ago-based industries also offer opportunities for field crops. Cassava roots, for example, are mostly used to produce animal feed in the form of pellets. Cassava can also be used to produce native starch from which a broad range of products can be made. One such possibility is to modify the characteristics of the native starch for use as an industrial starch that replaces potato, corn, and mung bean starch, etc. There are more than a hundred possible types of modified starch uses, such as filling and binding agents in paper, textiles, food, and adhesives. Another method is to convert the starch into various types of sweetener such as glucose, high fructose syrup, and also sorbitol, which is used in the cosmetics industry. Through fermentation, cassava can be used to produce monosodium glutamate, a widely used ingredient for seasoning, and lysine, an amino acid required in the production of animal feed. Cassava starch can be used to produce ethanol, and is also used to produce easily degradable plastics and environmentally friendly substitutes for plastic foam.

Similar opportunities may exist for other crops. For example, rice is used in the production of some beer in the US. After many refining processes, castor oil can be used as a lubricant for jet planes, etc. To date, these opportunities have been limited by the lack of knowledge of technological possibilities and sources. In addition, some food crops have become politicized, and price support policies and programs have made these crops unattractive as industrial inputs, because in order for this to occur prices need to be predictable and competitive and not determined by unpredictable political factors.

If the above opportunities are many in number, the next question is why they are not being taken up in Asia. The answer is twofold. First, the processing technology for many tropical raw materials other than food crops is not readily available in the international market. Second, government interventions related to some of the crops have become obstacles to increasing the number of value-added opportunities.

To understand the constraints placed on the development of agro-based industries, it is useful to classify agro-based industries into three types. The first relates to the processing of traditional raw materials that are internationally traded, for example the milling of rice and sugar. Technology is not a constraint because the production volume of these commodities is large enough and there are international technology markets in existence. The constraint is government interventions that attempt to increase prices above those of international markets, making it difficult for local industries to be competitive.

In addition, the government may also protect local growers by banning imports. This type of constraint is particularly severe when the commodity is a politically sensitive good, such as rice. Conversely, the government may be too eager to promote local processing and grant monopoly licenses. One example of this is castor oil production in Thailand, where imports of castor seeds were banned and licenses for primary processing were given to only one producer. No independent investors would want to invest in downstream industries where there is only one monopoly supplier. A reduction in government intervention is an effective way of promoting these types of industries.

The second type of agro-based industry includes those using raw materials that are tropical or local in origin, for example cassava and tropical fruits and vegetables. For these raw materials, there are no existing technology markets. The adaptation of existing technology to suit particular properties of local raw material is necessary. Science and technology for these industries will have to be developed locally. The public sector has a role to play in providing incentives and support for research, the dissemination of information about local inputs, and preliminary technology assessments, the output of which can be shared with interested investors for further feasibility studies.

The third type of agro-based industry is that connected with large-scale plantations introduced by processing industries. In this case, a system for the production of raw material specific to the industry has to be established or arranged. In contrast to the first type, the industrial processing methods determine the inputs produced. Examples are the production of palm oil, tuna canning, and pineapple juice. Investors are characteristically large enterprises or multinational corporations needing to organize the production of input on a much larger scale than found in the traditional methods of production. Although processing technologies like canning can be simple, substantial investment in advertising and marketing networks is required. In the past, most governments tended to favor this type of investment and to offer more incentives than required, despite the fact that investors in these industries are the most powerful of all. At the same time, governments have neglected actions needed to strengthen the industries in the first two groups.

When large-scale processing plants do not establish their own plantations or farms and have to rely on farmers as subcontractors, appropriate institutional arrangements to obtain reliable supplies of the appropriate quality are imperative. However, dealing with thousands of farmers is no easy task. Often, failures occur in factories where managers are unable or neglect to manage farmer subcontractors. In some cases, the government is also required to oversee the "fairness" of contractual arrangements or dispute settlements, e.g. in sugar and tobacco.

Apart from large-scale processing, niche opportunities also abound for small crops such as herbs, pot plants, natural dyestuffs, and various types of health food and organic products. Within Asia, high-income markets exist for exotic meats, foods, beverages, and food supplements such as marine eels, venison, and ginseng. In Thailand, women's groups from rural communities have been successful in placing organic cosmetic products, preserved foods, and chemical-free textiles and fabrics in supermarkets in large cities. In these areas, the role of the government is to help small entrepreneurs with phytosanitary regulations, packaging technology, and international marketing expertise in order to upgrade their operations so that they can compete in international markets.

Biotechnology

Biotechnology is widely believed to be the driving force that will provide the next major increases in agricultural productivity. It is also expected to help lessen some of the adverse impact of agriculture on the environment, especially that caused by pesticide pollution.

In reviewing the biotechnology products that have come to market, or are likely to do so in the short to medium term (5–10 years), there are none so far that are likely to increase potential crop productivity on the scale of the green revolution. Apomixis, as indicated earlier, is a possible biotechnological tool now under study at some of the international research centers. If it can be successfully transferred to major crops like rice and wheat, yields could increase by 20–30 percent through heterosis. The real impact of apomixis would be to reduce the cost of hybrid rice and wheat seed.

Benefits from biotechnology are already evident in the lessening of impact from crop production on the environment, in decreasing pollution, and in increasing the efficiency of pesticide use. Herbicide resistance created by biotechnology is expected to lessen the impact from herbicides by decreasing the amounts used as well as making it possible to use chemicals with milder effects on the environment. The use of pest-, disease-, and herbicide-resistant varieties is also expected to decrease production costs by reducing the cost of pest control as well as preventing yield losses.

Resistance created by biotechnology will, however, be just as prone to breaking down as resistance created by conventional breeding, as happened in rice with resistance to the brown planthopper and in wheat with rust pathogens. A number of insects have already been identified in the US as resistant to the Bt toxins (greater detail is provided in a companion volume, Ammar (1999)), which are produced by new crop varieties that have had genes inserted from the bacteria *Bacillus thuringiensis* (Bt).

The application of biotechnology to crop production has so far concentrated on crop protection, i.e. from insect pests, pathogens, and herbicides. These have involved the transfer of various resistance and tolerance mechanisms, mainly from bacteria, into plants by employing recombinant DNA technology. Although they originated in the US, many of these new crop varieties (of maize, cotton, soybean, etc.) are already being adapted to and/or field-tested in Asia through the private R&D system of multinational agribusiness companies, often in close collaboration with national (public) agricultural research systems. Commercialization is, however, being held back by the fear that it would be difficult to stop farmers from keeping some seed for their own use in subsequent years or even selling it to others. In order to overcome this, a new system in which the second generation of seed is aborted, is now being explored in the seed industry. The system, termed "terminator technology" by NGOs, will enable farmers to produce grain for sale from the seed but make it impossible for them to use some of the harvest for seed for the next season.

Commercial interests dominate agricultural biotechnology R&D. Internationally, these are represented by a small number of multinational companies. Even in developing countries such as India with the strength to develop biotechnology programs, research emphasis is often placed on export crops, and not on the basic staples of rice and wheat. There is a concern that this will make developing countries increasingly dependent on industrialized countries for inputs (Hobbelink, 1991, cited in WRI, 1994, p. 6).

More significant, however, is the potential for biotechnology to contribute to the widening of the gap between the rich and the poor. This technology, as well as the science that supported the green revolution, was the result of publicly funded R&D; benefits from technology transfers to countries in Asia were basically free of charge. The green revolution, therefore, helped to close the gap between the rich and the poor to a significant extent. Even then, many have been critical of the green revolution by claiming that it bypassed many who live in LFEs. Biotechnology will leave even more people behind. Some countries simply cannot afford the heavy investment that biotechnology requires. In some areas and crops, rice and wheat among them, the profit incentives will not be sufficiently attractive to private investors. The Indian seed industry, which is currently aggressively driving the diffusion of hybrid maize in India, has so far stayed out of Rajasthan, Uttar Pradesh, and Madhya Pradesh, where maize is grown as a food crop with a very low input level (Morris, Singh, and Pal, 1998). Even in the US, private R&D on self-fertilized crops, such as wheat and soybean, is relatively limited, especially following the enactment of the Plant Breeder's Protection legislation with the "farmer's exemption" clause (i.e. enabling farmers to keep seed for their own use).

Countries without biotechnology capacity will increasingly see their traditional exports displaced by substitutes derived from biotechnology. For example, rapeseed plants with more than 35 percent laurate in their oil have been produced by Laurical (Calgene, LLC), and are now marketed in the US. The new plants are expected to provide a cheaper alternative to coconut and palm kernel oil. Such losses in competitive advantage will take place not only between developed and developing countries, but also between developing countries with strong biotechnology capacity and those without.

Theoretically, the adverse effect of such displacements on the loser should be temporary only. Adjustments would occur over the long term. However, it would seem that those countries with a limited biotechnology capacity would find themselves with increasingly limited options in making adjustments.

Several international initiatives have been mounted in order to share the benefits of biotechnology with developing countries. For example, the Rockefeller Foundation's Rice Biotechnology Network started in 1985, with the aim of improving the biotechnology capacity of developing countries, mainly in Asia. There is also increasing collaboration between private and public sector R&D. IRRI, for example, is cooperating with two private companies in its attempt to transfer the Bt gene into rice. In one case, IRRI paid a fee to Plantech of Japan to use its Bt gene for research purposes and has the option to buy the gene outright. Another case involves a Bt gene that has been provided free of charge to IRRI by Ciba-Geigy of Switzerland. Bt rice from IRRI will be made freely available throughout the developing world, but not to Australia, Canada, Japan, New Zealand, the US, and members of the European Patent Convention.

In addition to its potential impact on crop production, biotechnology is also causing major changes in the organization of agricultural R&D. Innovations in biotechnology and its related and new discipline of genomics (the molecular characterization of species) has forced a broadening of the scope of plant science R&D.

In conclusion, biotechnology definitely holds many opportunities for the private sector. For the public sector, it offers great opportunities in facilitating R&D. Incentives for biotechnology production in the private sector will have to come from temporary monopoly, as granted by patents or plant breeders' rights laws. Thus, governments must ensure that small farmers will not be in the position where they become locked into a no-option situation.

V ASIAN AGRICULTURE: TOWARDS 2010

Asia has two distinct cropping systems, one for favorable environments, i.e. those that are irrigated or have reliable rainfall, and one for less favorable environments (LFEs). Over the last two decades they have developed along very different courses, with cropping systems for favorable environments receiving far more attention and resources, and being significantly improved as a result. This trend of improving cropping systems for favorable environments relative to those for less favorable environments will continue to 2010.

In favorable environments, crop yields will continue to grow over the next 12 to 15 years. However, it is not simply yields that will determine whether or not there will be a real decline in area sown of the two major staples, rice and wheat. As there are only a few places in Asia where farmers grow rice or wheat for lack of options, profitability has been playing an ever-increasing role in crop choice. As a consequence, trends towards greater crop diversification will continue. Therefore, increases in the yield potential of rice and wheat will not guarantee growth in overall production. Furthermore, improvements in production technology, including those derived from biotechnology, will only be translated into cheaper food for the poor when they significantly lower the cost of production. The role of the private sector in agricultural R&D will continue to expand. It will, however, be restricted to specific crops, especially those with hybrids like maize, rapeseed, sunflowers, and vegetables, and is unlikely to be extended to major field crops such as rice, wheat, soybean, pulses, and roots.

Asia's LFEs come in many shapes and forms: areas with unreliable rainfall coupled with no access to irrigation, land with poor soil, irrigated land suffering from salinization, steep slopes prone to erosion, and dry land threatened with desertification. The increased productivity of cropping systems in favorable environments brought about by the technologies and policies behind the green revolution has indirectly benefited some inhabitants of LFEs by increasing their employment opportunities. Most inhabitants of LFEs, however, have simply fallen further behind, and in areas where access is difficult, food security remains a major concern.

BUSINESS-AS-USUAL SCENARIO

Under the business-as-usual scenario, high growth rates in the yields of food crops, especially rice, will not be sustainable. This is due to the high costs associated with maintaining growth and the continuing diversification out of food crops and into crops with higher profit margins. The high costs associated with maintaining growth are the direct results of poor management and environmental degradation. If the current institutional management policies remain unchanged, the lessening availability of water will increasingly constrain the potential for sustainable growth. As the competition for water between different users increases, water will inevitably be shifted away from agriculture, where the marginal product of water is relatively low. A simulation study conducted at the International Food Policy Research Institute (Rosegrant and Ringler, 1988) revealed that this would have a substantial impact on the global food supply. Both yield growth and crop area growth would slow down. The impact would be particularly strong on rice, as it is a water-intensive crop. The average price of rice would increase by 68 percent between 1993 and 2020, leading to a large increase in malnutrition. The study assumed that irrigated land would be lost because of land degradation, urban encroachment, and loss of water for irrigation as water is increasingly used in nonfarm activities.

The coastal, aquatic, and wilderness resources that are essential food items and contribute to the livelihood of the poor will be mostly depleted. Furthermore, more farmers in resource-poor and coastal areas will be left further and further behind as the world continues to move towards the electronic age.

VISION 2010

A desirable vision for 2010 is one considerably different from the situation described above. It is one in which Asia is free from hunger and Asian agriculture has an increased and sustainable capacity for more equitable and greener growth. Green growth is that in which increases in productivity do not arise as a result of the unsustainable use of natural capital and the environment. Production increases should come about through higher yields per unit area, and not simply through increasing the amount of land under cultivation. Equitable growth is growth whose benefits reach all parts of the community, even the poorest of the poor. As food security on the Asian continent becomes a lesser concern at least up until 2010, more attention needs to be paid to the elimination of malnutrition.

Towards the year 2010, farmers in favorable environments and in advanced agricultural areas will enjoy a wide variety of options for crop mix and technology. This should help them reduce their use of natural resources and fossil fuels. The farmers of the future will form a heterogeneous group that contains a diverse range of interests and skills. This diversity results from the different environmental conditions faced by farmers and will be augmented by broadened crop choices and new market niches. Some farmers will engage in high-input, high-output intensive farming and some in ecological farming. Thanks to improving communications infrastructure, information about changes in technology and the market will be readily available to farmers at a low cost. Computerized farming aiming at optimum use of input (especially in horticulture and aquaculture) will co-exist alongside more conventional methods. An increased level of farmer-to-farmer knowledge transfer will be made possible through distance media and electronic mail. Increased awareness will help farmers of the future to be constantly aware of new production technologies and enable them to select those that minimize impact on their health as well as that of consumers and the environment, helping to guarantee the sustainability of their production systems.

More technologies will be available to farmers in LFEs by 2010, enhancing their productivity and the productive capacity of their land and natural capital. Incentives would have to be provided to encourage farmers who need to make additional investments to minimize their impact on third parties or society as a whole. Increased opportunities in education and nonfarm means of generating income, e.g. ecotourism, should be made possible with improved infrastructure, particularly roads, access to mass media, and clean water.

The farming communities and groups will themselves have more control and influence on the use and maintenance of the natural resources and public infrastructure related to agriculture, as well as over the direction of agricultural R&D. This increase in control will be proportional to their increased willingness to share in the cost of maintaining local public goods and effective R&D. As a result, those local institutions that prove themselves effective in the provision of local public goods, the conservation of local common areas, and the allocation of user rights will become strengthened and duly recognized by law.

The efforts of national and the international scientific communities will continue to provide wider and deeper technological options, and to enhance the sustainable use of natural resources. Research efforts to prevent yields from declining, maintenance research, and attempts to increase yields will continue. Their approach, however, will have been modified, as research targets become more focused on farmers and specific locations, with a greater emphasis being placed on LFEs and new innovations for resource-poor regions. Natural resources and the environment will become other important objectives for agricultural R&D. Publicly funded international and national agencies should have sufficient resources to conduct RD&E on pest control and the use of biotechnology to develop pest-resistant and mineral-efficient varieties of crops. Improvements in the efficiency of water usage will be a new addition to the research agenda. National research capabilities will require strengthening in those locations, the semi-arid and humid tropics for example, most likely to feel the effects of climate change. An equitable international system regulating the exchange of genetic resources needs to be instituted.

Civil society organizations should have acquired scientific and technological knowledge and combined this with practical field and social skills in order to achieve the twin goals of improved production and conservation. The advantage they currently possess in participatory processes could be expanded to serve the purpose of increasing the responsiveness of the public RD&E apparatus to real onsite needs. Civil society organizations will play active roles in the dissemination of technology, becoming an effective link between farmers and the public sector.

The market will continue to be a major driving force supporting agricultural development. Private-sector involvement in agriculture will continue to grow, not only in its current role in the seed and chemical industries, but into new roles in R&D, including biotechnology, mechanization, irrigation, and extension services. The fertilizer and crop protection industries will play an active role in the promotion of integrated nutrient and pest management.

Agriculture will no longer be mistakenly perceived as a sunset industry, but a vibrant life- and growth-support system. Agriculture will be viewed as a sector that offers income-generating and employment opportunities, and not as the sector of last resort.

In order for these visions to be realized, the main priority for governments will be to strengthen their policies and institutions regarding natural resources. The overextraction of open-access resources and multiple-use conflicts will have to be resolved through a combination of economic, legal, and social instruments. Responsible agriculture and fishing will need to become the prevailing code of conduct. Conservation of natural resources and the environment has to become an additional national objective, and biophysical planning will have to be the norm rather than the exception. The public sector must continue actively to support investments in education and technology. Large infrastructure investments need to be based on rigorous cost-benefit analyses, with due recognition being given to social and environmental costs and payoffs. The public sector, however, will not necessarily continue to be a direct provider of public goods, but rather promote and facilitate private investment and adopt the role of regulator to provide a level playing field for private operations, from individual farmers to multinational corporations. Government operations will become less labor intensive, as the labor-intensive activities of maintenance and monitoring become privatized or devolved to local organizations. The public agricultural agencies will have to adopt a more flexible, adaptable managerial and catalytic role.

VI CONCLUSIONS AND RECOMMENDATIONS

CONCLUSIONS

To date, the green revolution has enabled increased food production through a package based on HYVs that are responsive to fertilizers and good water control. However, an investigation into production trends in Asia has found that the rates of production growth and yield growth of major food grains are showing declining trends. These trends are most obvious for rice, Asia's most important staple crop. This does not mean that the potential for further growth is necessarily exhausted. Wheat and maize still have a substantial capacity for further productivity gains. Their yield growth rates are still robust even after a recent slow down. For these three crops, considerable potential also exists for productivity gains to be achieved through increased efficiencies from improved crop management.

Increases in yield potential, however, may not necessarily translate into yield growth unless they are accompanied by an increase in net returns. Profitability or net returns from food crops has dwindled over time, driving farmers to alternative crops with higher margins, such as oil crops, fruits, vegetables, and sugar cane. Although this may not affect the overall productivity of Asia's cropping systems, it will certainly lead to a decline in food grain production.

In the livestock sector, especially the monogastric sector, production has been market driven. However, growth has been particularly rapid in urban centers and has resulted in an intense concentration of production units in peri-urban centers, increasing pollution and health risks to the extent that long-term growth may not be sustainable. There is a role here for the public sector to coordinate production so that waste discharges can be recycled as an energy source or be put to more efficient uses.

The outlook is bleak for marine fisheries. Although growth has been strong, the long-term sustainability, especially that of coastal fisheries, is greatly threatened by overfishing and pollution. Coastal and inland aquaculture is threatened by pollution from outside sources. Coastal aquaculture, especially shrimp farming, may itself undermine the sustainability of other agricultural systems if not properly managed. In addition, inland aquaculture is constrained by the limited availability of water of suitable quality.

Environmental degradation related to agriculture is a product of technological and policy failures. High-input technology creates onsite second-generation effects, but they can be corrected by improved RD&E. In LFEs, the lack of appropriate technology is a major source of environmental degradation. A lack of appropriate policies and institutions and lax law enforcement are the main sources of external costs and the wasteful use of resources.

Second-generation problems related to the high-input technology package, as well as negative impact on human health and the environment, are also perceived as being detrimental to future growth. Intensification-induced declines in productivity growth have been suggested as a possible threat to future growth in crop production. Many of these problems, however, can be solved by improving field-level knowledge, better crop management, and better communication between farmers and R&D officials. The achievement of sustainable agriculture will also require that the current mode of crop-based and laboratory-oriented R&D is adapted to field- and farmer-based technology transfer systems. Agricultural R&D and technology transfer will have to be sufficiently adaptive and responsive to deal effectively with these problems as they arise. This becomes even more challenging when dealing with agriculture in LFEs, which have only marginally benefited from the green revolution to date.

The sustainability of Asian agriculture will also depend on the prudent use of natural resources and careful consideration for the environment. The natural resource base of Asia is now under great stress, and this will become even greater as the population continues to increase. Investment in environmentally sensitive technology is needed to ensure sustainability. The current constraints related to natural resources are not the results of limits in supply but rather are managerial and institutional problems. The solutions to the current problems in sustainable agriculture no longer simply lie in technology, but also in institutional reform.

Sectoral policies, especially policies related to natural resources, are outdated and lag behind the socioeconomic changes that have altered the patterns of resource use. For example, throughout Asia water-resource management has been fragmented and project based. Both surface water and groundwater are mostly under open-access regimes that encourage wasteful usage, which in turn may lead to waterlogging and salinity problems. Water pricing has been adopted by many Asian countries, but mainly for the purpose of paying for the operation and maintenance costs of irrigation only, rather than as a basis for allocation purposes. Removing policy distortions and institutional constraints in the natural resource sector, while at the same time promoting participatory management, is key to developing the long-term sustainability of both the agricultural and the agri-based sectors.

Some of Asia's crop production is on fragile land. The mismanagement of fragile lands leads to rapid degradation (e.g. soil erosion, salinization, waterlogging, desertification) and is often not just due to simple mistakes of farmers, but is symptomatic of a set of complex social, economic, and ecological problems. Failures of national policies, trade, and investment, as well as sectoral regulations such as on soil erosion control, have all contributed to the degradation of natural resources and the environment. In the past, except for the socialist countries that have adopted market-based reforms, most Asian governments opted for technical solutions. These are only a partial answer to the problems; policy and institutional reforms are necessary to tackle the problems in their entirety. In addition, the issue of poverty has not been appropriately addressed. Rather, it has been used as an excuse for handout policies or to implement price guarantee projects designed to win political support.

In the past, technology was used to circumvent the need for reforms that may have been economically and socially desirable but politically impractical. In the future, appropriately designed technology will remain a very important tool, but it cannot solve all the problems and sometimes creates problems of its own, especially when misused. More importantly, the green revolution has ignored LFEs, which make up a large part of agriculture in Asia. A wider and deeper understanding of the complex relationships between nature, technology, and institutions is necessary. For example, it must be remembered that a discovery of a sustainable agricultural cropping system does not by itself guarantee sustainable agriculture. It is sustained good governance that will result in performance that meets economically and socially desirable objectives.

STRATEGIES FOR SUSTAINABLE AGRICULTURE

Outlined here are prioritized strategies and sectors needed to achieve the long-term vision of greener growth and a hunger-free Asia, together with the necessary policy and institutional reforms to implement them. Other recommendations, mentioned or implicit in the earlier chapters, are included in Annex B. The policy and institutional reforms that are also necessary for the effective implementation of the proposed strategies are discussed. Three strategies are presented. They highlight the adjustments needed in the current directions of agricultural development. The first strategy calls for sustained support for investment in agricultural technology, requiring adjustments in the objectives and methods for both high potential areas and LFEs. The second strategy focuses on LFEs and the need to incorporate institutional considerations, and on immediate ecosystem concerns. The third strategy highlights comprehensive river basin management, which will optimize both production and conservation objectives.

Strategy for Integrative Technology Production and Transfer

The analyses in this volume suggest that a new push for increased productivity, and hence a sustained investment in agricultural research, are necessary. This is especially so for rice for which there are early warning signs of weakening sustainability. This is important for policymakers in international agencies and national governments to recognize, particularly in the PRC where expenditures on agricultural research have declined. However, the increased and sustained support required will be effective, and will achieve both increased productivity and sustainability, only if the existing system of technology production and transfer is modified.

Much of the new growth will have to come from increasing the efficiency of the cropping systems, where there are still considerable opportunities for further productivity gains. The causes of existing inefficiencies are often complex, and cannot be overcome by simple, broad-based solutions (i.e. ready recipes) prescribed in a top-down method in the manner of the widely adopted HYVs.

To achieve sustainable development, a three-pronged approach to agricultural technology production and technology transfer is recommended: management for sustainable agriculture needs to be 1) oriented around natural resources and the environment (NRE), 2) participatory, and 3) based on science. Putting NRE objectives at the fore does not mean that output maximization is no longer an objective. Output must be maximized but full recognition needs to be given to NRE constraints and consequences. Participation means that there must be two-way communication between extension workers and farmers, and between production-oriented and conservation-oriented agencies. Local knowledge and social capital must be harnessed. This does not mean that the agricultural system will be less scientific in its approach. In fact, it means that the system will be more science and technology based, bringing science to the fields and adapting it to better benefit local users. It means that successful systems will be less centered around a particular crop and more oriented to particular locations.

Asia's most productive land is already being intensively cropped. There is usually more than one or even two crops per year. High inputs are used that provide high yields, which results in high rates of nutrient removal from the soil. New problems that will threaten sustainability are inevitable. Location-specific solutions will have to be devised for each situation. Considerable gaps remain between the yields produced at experimental stations and those produced on farms located in favorable environments. There is already a readily discernable yield gap separating the more favored and the less favored environments. More emphasis will have to be given to crop management R&D and biotechnology. This will require capacity building for local research facilities and development. Environmental impact assessments should be included as part of the technology assessment, as well as for management practices such as crop management.

The above suggestions require nothing less than institutional innovations that will incorporate the all-important feedback mechanisms between the technological innovators and the technology users. Future development strategies will have to emphasize knowledge-based production systems that focus on users rather than researchers. The experience of agricultural extension in the PRC could be emulated and adapted. Productivity increases can be enhanced by fine-tuning activities rather than by large-scale public investment programs.

Research objectives will have to focus more on cost aspects that emphasize a reduction in the use of chemicals and fossil fuels, as well as of renewable natural resources. New plant breeding innovations and biotechnology offer opportunities for sustained increases in yield and for prevention of crop losses from pests and diseases, but require input from existing genetic resources and a free flow of genetic material.

The current multilateral genetic exchange system has functioned relatively well, although the US Patent for basmati rice lines and grains has created a feeling of mistrust and unfairness among providers of genetic resources.   An international management system or code of conduct that recognizes and protects traditional or prior users' rights while providing sufficient incentive for private R&D initiatives is required in order to maintain the free flow of genetic material to international research centers.

In place of a simple plant-breeding objective, such as doubling the yield potential by modifying the plant type, crop breeding programs now have to be concerned with a diverse range of issues related to gene management.   In order that these issues are addressed in an integrated manner, the CGIAR's Third System Review (CGIAR, 1998, p. 26-27) has recommended an "integrated gene management" approach as a basis for activities in international agricultural research centers (IARCs) and national agricultural research systems (NARS), which includes:

- patenting processes for new varieties, and placing their use under free licensing;
- a legal entity that could hold CGIAR patents;
- the conservation of agrobiodiversity and its sustainable and equitable use;
- research on genomics and molecular breeding for the purpose of supporting NARS to enhance the productivity of major farming systems in an ecologically, economically, and socially sustainable manner;
- strict adherence to the equity and biosafety provisions of the Convention on Biological Diversity and national government regulations;

- a central coordinating and servicing unit for advising both IARCs and appropriate NARS;
- a widened food security basket through the inclusion of minor and under-used millets, legumes, tubers, and other crops;
- the use of Mendelian and molecular methods of breeding in an integrated manner;
- an effective public information and communication system, with total transparency and accountability for work in the field of biotechnology; and
- a (CGIAR) system-wide review of plant breeding efforts, with the aim of freeing up resources for new priorities while accelerating the introduction of modern marker-assisted breeding and bioengineering technologies.

The NARS will have to face most of these issues at the national as well as the international level. This expansion in scope means increased demands for research funding and personnel. As recommended for the CGIAR System, there will be attempts to "free up" resources in NARS in order to establish new priorities. Asia-wide discussions and debates on these priorities would be very useful, especially for the smaller NARS.

In addition, biotechnology and genomics will now play a major role in the future growth of Asian agriculture. A considerable proportion of the region's R&D resources is now being redirected to build biotechnology capacity at the expense of research in other areas. The potential of biotechnology, however, cannot be realized without understanding the genetics controlling important traits. Mechanistic explanations of how certain traits are expressed, especially quantitative traits such as yield, nutrient efficiency, and tolerance to drought, acidity, and salinity, will be essential for the identification of major genes. Asia-wide collaboration and networking in these essential research areas could create significant savings. A sustained level of increased investment in R&D by the public sector along with international support is necessary in areas where biotechnology can benefit small farmers and resource-poor regions. Otherwise, the fruits of biotechnology research will only be available to wealthy farmers and private corporations, further aggravating income distribution inequities.

Strategy for Less Favorable and Fragile Ecosystems

Asia's less productive cropland has been bypassed by the green revolution. The LFEs are not homogeneous in terms of their characteristics, and include poor and degraded lands in highlands, uplands, and lowlands, with each area having its own particular difficulties. Public policy on improving the livelihood of people in these areas needs to be based on assessments of investment costs and potential returns for each particular location. The first step towards the sustainable development of LFEs would be to classify them according to investment potential. It is imperative, however, that social and environmental goals as well as economic goals be considered in such classification.

In addition to improving traditional food crops, such as cereals and pulses, new development activities might include alternative products such as nuts and palms, berries, fruit, wild game, tree crops, livestock, and fishery activities. Nonfarm and off-farm activities such as ecotourism, agro-processing and manufacturing are also possibilities. Another essential element of any LFE policy is that it should be flexible enough to be able to adopt new institutional and technological changes as these become available, and to be able to respond to new problems as they emerge.

Many of the new innovations that will be necessary for increasing the productivity of LFEs will be highly location specific. Local capacity building is therefore an indispensable component of any effective RD&E effort. Asia-wide R&D, however, still holds some of the most promising returns to public investment for certain major food crops, such as rice, wheat, and grain legumes. This research remains vital because these crops will continue to be the most important production activities in many of Asia's LFEs. It is possible to breed crop varieties that are tolerant of or adapted to conditions in LFEs, and such solutions would help to increase productivity to a certain extent.

Recent progress in plant breeding is already demonstrating promise for improved rice cultivation in rainfed lowlands (through drought tolerance) and in flood-prone areas (through tolerance to submersion) in many countries. New breakthroughs in breeding for nutrient efficiency, especially for phosphorous in rice and soybean and boron for wheat and pulses, are also occurring. These will not only decrease production costs (by increasing yields while reducing fertilizer use) but also eliminate the need to transfer complicated fertilizer management technology.

The case for breeding boron-efficient wheat provides an example of how potential crop improvements in LFEs might be addressed. Boron deficiency is a real and widespread limitation to wheat production. It can lead to 100 percent yield loss in, e.g. Bangladesh, the southwestern provinces of the PRC, the northeastern states of India, Nepal, and possibly also areas of Myanmar. Farmers with boron-deficient soil are also prevented from adopting newer varieties that are higher yielding and disease resistant but that are more susceptible to boron deficiency. A potential solution has already been identified. Boron efficiency has been found to be a genetically controlled trait in wheat; more efficient varieties can deliver 100 percent grain yield under the same conditions that prevent less efficient varieties from giving any grain yield at all. Also, boron deficiency is a regional problem, which provides the opportunity for economies of scale through an Asia-wide R&D program. The total area affected is some 2–3 million ha, located in small wheat growing countries whose technical capacity is limited (Bangladesh, Nepal, and Myanmar) and in marginal areas in larger countries with greater technical capacity (India and China).

For many situations in the LFEs of Asia and especially in fragile ecosystems, tolerant varieties will only be a small part of the solution. Such situations are generally characterized by one or more of the following conditions: (a) the social and economic circumstances of farmers as well as the physical conditions demand that technical solutions be tailor-made for each specific case; (b) social and institutional solutions are essential, in addition to and in conjunction with, or instead of, technical solutions; (c) problems require management at a level beyond that of individual farms. Attempts to increase the productivity of cropping systems in a sustainable manner will call for an integrated approach to natural resource management (INRM).

INRM has three important elements: (a) a holistic focus on the entire ecosystem rather than on individual fields; (b) farmer participation in the R&D process; and (c) recognition that social/institutional solutions are often required in conjunction with, in addition to, or instead of technical solutions.

To increase the productivity of cropping systems in irrigated lands with salinity/waterlogging problems (PRC, Central Asia, India, Pakistan), salt-tolerant varieties of crops (wheat, for example) can make significant contributions. Salt-tolerant varieties alone, however, will not be enough. The management of the water table is an essential element of the management of land with both salinity and waterlogging problems. The water table has to be managed on the basis of the catchment, which may or may not fall within the boundary of individual farms. Where the catchment is within a single farm, the farmer will still have to manage on a farm-wide basis, not on a field-by-field basis. The areas most affected by salinity, that are too saline for even the most tolerant varieties of wheat or other grains, should be set apart for salt-tolerant fodder species such as Kalar grass or saltbush. Salt-tolerant deep-rooting trees (e.g. eucalyptus) may have to be planted at strategic locations to draw down the water table. Collaboration between neighbors for water table management will be essential where one catchment covers many farms. In addition to the availability of salt-tolerant varieties of crops, fodder species, and deep-rooting trees, technical knowledge such as the identification of the catchment boundary (which does not always follow the external contours of the land) as well as social organizations that could facilitate collaboration among neighboring farmers would be required. The ecosystem under consideration in such a case would be on the scale of the individual catchment, in which the water table has to be managed, and farmers' participation would directly determine the form of collaboration through which water is to be regulated.

The sustainable management of cropping systems on steep slopes prone to erosion in the uplands and highlands requires not only an increase in crop productivity, but also a minimization of adverse offsite effects on those living in the lowlands and society at large through the various services provided by the uplands/highlands. These services range from the regulation of the water supply (from watersheds), the control of wild/forest fires, the conservation of forests and biological diversity, to carbon sequestration and the prevention of siltation in rivers, reservoirs, waterways, and irrigation canals in the lowlands.

The measurement of, for example, stream siltation, forest cover, biodiversity, stream flow (amount and seasonal distribution), and the incidence of forest fires, would provide quantitative indicators with which the success or failure of management could be judged. It is, however, essential that an appropriate set of such criteria be made available to each level of management, whether it be at the level of the farm, community, catchment, or watershed. Where cropping systems are sufficiently productive (e.g. hybrid maize, high-value vegetables, fruits, and flowers), the cost of minimizing the various adverse impacts may not be overly burdensome. However, in general, cropping systems in fragile ecosystems produce barely enough to feed the local population; any expectations that the latter could also work to save the environment and natural resources would be unrealistic. It is essential to all INRM projects characterized by major offsite impact that farmers' contributions and trade-offs to environmental conservation be fully recognized.

There are already some basic innovations in crop and land management that have proven successful throughout Asia in improving the performance of cropping systems on steep slopes. These include land allocation according to grade (degree of slope) for different types of cropping systems (according to their potential to cause soil loss), e.g. rice on flat lands with water, upland crops on milder slopes, and woody perennials on steeper slopes. There are also various erosion control measures, including some that are relatively low cost such as contour vegetation strips. Crop management may benefit from genetic improvement such as traits for tolerance (e.g. to acidity, disease) and efficiency in use of major limiting nutrients (e.g. phosphorus). However, in contrast with the uniformity of conditions required by green-revolution technology, the diverse conditions in LFEs and fragile ecosystems require that technology be adapted to local conditions for its effective transfer, and a prerequisite for adapting R&D is farmer participation.

It is a widely held view that investments in LFEs tend to reap lower returns than do similar investments in favorable environments. Investigations of soil and climate conditions of public investments in India in 20 agro-ecological zones over the period 1970 to1994 suggest that the contrary may be true (Hazell and Fan, 1998). The low-potential rainfed systems demonstrated the highest marginal returns to production (measured in rupees per unit input) when compared with investments in canal irrigation, roads, market developments, and education in irrigated and high-potential rainfed areas. The returns from adoption of HYVs to low-potential areas were almost as high as those to high-potential rainfed areas, and both earned higher returns than did investments in irrigated areas. These very favorable results are believed to be a consequence of spillover effects.

Strategy for Natural Resources and Environmental Management for Sustainable Agriculture

In the long term, natural resources and the environment are necessary components of all the three priority strategies for the achievement of sustainable agriculture. The difference is that in the first strategy, natural resources and the environment are included as one of the objectives of R&D, with the expectation that new technological packages involving onfarm practices will minimize both onfarm and off-farm environmental impact, and that technological innovations will be environment enhancing. In the second strategy, the technological management package addresses management at the landscape level, for example small watershed management. It also requires local institutional support recognized by law. In the third strategy, management takes place at a broader scale, e.g. at a bioregional level such as a river basin. This level of management addresses cumulative impact within and between sub-basins. Using this approach, the interaction of different resource uses is taken into account, which renders trade-offs more transparent. An example of large-scale bioregional planning is the development plan for the Mekong River basin under which transboundary impact can be managed.

Achieving the aims of the third strategy requires a longer time frame than do the other two strategies, and a few preparatory steps are necessary. First, the present system, which is based on administrative boundaries, must be readjusted to one based on biophysical or bioregional boundaries for the purpose of gathering information related to natural resources and local environments, their interrelationships, and interactions. Second, the identification of critical areas is necessary (Khan, 1996). Critical areas are of two types: those important to long-term agricultural sustainability, e.g. spawning grounds, biodiversity-rich habitats, and fragile ecosystems where potential degradation and multiple-use conflicts are imminent; and those of high growth where sustainability indicators are showing early warning signs of degradation. Once this information is in place, planning at the bioregional level can proceed. The planning process has to encompass simultaneously economic, social, and environmental considerations.

Presently, environmental planning is often a stand-alone process with the ministry of environment acting as the sole protector of natural resources and environment. Under the proposed strategy, growth and sustainability issues, and the corresponding growth-oriented development and conservation projects, are juxtaposed, prioritized, selected, and scheduled. Environmental and social impact assessments also have to be undertaken at the planning level, prior to implementation.

Using this approach, the management of natural resources will occur at the bioregional level, for example by river basin committees consisting of representatives from sub-basins. The organization should be bottom up, i.e. starting with sub-basin committees from the lowest level. Each country should start with the region of highest economic and environmental priority, or with highest level of multiple-use conflicts.

This approach, as proposed, would comprise participatory planning, the establishment of principles for the allocation and use of natural resources and their management, zoning, and development of land-use plans. The issues relating to rights to the use of natural resources and the protection of these rights would have to be specified and established.

Wherever the capacity for effective local government and social organizations exists, the devolution of some responsibilities, e.g. local water resources, fire protection, and community forest management, has proven efficient and effective for both allocation (in the case of water) and conservation practices. The principle of the devolution of rights to and increased responsibilities for local communities and governments for different resources would have to be specified, acknowledged, and legalized.

Where local organizations and social capital do not exist, the identification of existing constraints and capability building are necessary for the achievement of long-term growth. In either case, a check-and-balance system from the central government continues to be necessary in order to assure transparency and accountability at the local level.

POLICY AND INSTITUTIONAL REQUIREMENTS

The natural resources that support agriculture will not be adequately conserved if they are undervalued. Environmentally friendly technology, such as biogas energy, will not be adopted if the use of fossil fuels continues to be subsidized. Investment in conservation practices will not be worthwhile if land ownership is not suitably defined. A package of necessary policy requirements and reforms is therefore recommended.

Agricultural RD&E Policy

In order to combat hunger, maintain productivity in irrigated areas, and raise productive capacity in the LFEs, more investment is needed in R&D for agriculture, with the main priority being the raising of yield ceilings for rice. LFEs have to be the acknowledged target area for productivity improvements. The main priority in LFEs is the acid sulfate soil ecosystem. A national consensus on priorities may have to be developed through the media and through discussion.

The current top down RD&E system will have to be reversed to one that starts locally, with commensurate funding increases to local agencies. A farmer-focused RD&E system will have to be designed at the district level. Local agricultural colleges could be drawn into collaboration with the local RD&E system. Rewards for scientists and extension officers would be based on the field performances experienced by farmers. R&D funding may not be limited to public agencies but could be extended to learning institutes, NGOs, and private companies on a competitive basis.

A pilot project, in which extension activities are open to competition between the private sector, NGOs, and relevant government agencies, could be undertaken in agriculturally advanced areas. It would be under the supervision of local governments, farmer cooperatives, or water users' associations, as appropriate. In such a situation, a block grant may be provided to the implementing organization. In the longer run, the contributions of farmers to the extension system would gradually assume more importance than government grants. The willingness of farmers to pay would also serve to measure the value of the extension system.

Continuous capacity building is one of the indispensable components of an effective RD&E system. Scientific staff need to upgrade their skills constantly to keep up with international progress. Distance education and extension through television and radio could be introduced for junior, senior, and female farmers. Farmer-to-farmer transfers could also broaden the perspectives of not only other farmers but also extension officers. Integrating scientific knowledge with traditional and indigenous wisdom would have to be promoted through innovative means, for example district competitions of agrobiodiversity of genetic sources.

Natural Resources and Environmental Policy

Natural resource policies tend to be among the most outdated policies of many developing Asian countries. The first priority for reform is to reflect fully the scarcity value of natural resources in costs to users. This includes the value of natural resources both as inputs and as sinks. In other words, the open-access regimes that prevail despite resource scarcity will have to give way to systems where resources are properly valued and priced. Costs of such activities as pollution, which are currently external, will have to be internalized.

In Asia, two natural resource sectors that are priority sectors for reform are water and coastal and ocean resources. At present, the instruments used to correct for market failures in these sectors are mainly legal and regulatory instruments, implemented under command-and-control regimes. The continuous deterioration of natural resources and the environment to date demonstrates that these regimes are no longer effective in achieving both growth and sustainability objectives concurrently.

Other instruments such as economic instruments, concessions and property rights, pricing, charges, fees, and transferable development rights, need to be employed appropriately (see below under priority sectors). Social instruments can also be very useful in attracting public attention to sustainability issues and, especially where voting is important, in creating the grassroots demand for reform that is essential for affecting reforms. The first step is to promote public demand for good governance, without which development efforts often prove costly and futile. To this end, educational institutions, mass media, and social organizations provide important means for communicating with the public. However, at present, education and information on sustainability tend to lack focus and an adequate scientific basis; social organizations need both scientific and financial support from the government in order to undertake these activities.

Other Policies

Equally important is the need to remove price distortions created by other policies that favor the use of environmentally unfavorable practices, for example, State subsidies on fertilizers, and subsidies for agrochemicals, fossil fuels, and electricity. The removal of these distortions would decrease the wasteful use of resources, encourage the use of greener energy, and increase the incentives for using integrated pest management, soil and water conservation, and the search for alternative technologies. Similarly, distortions in output prices through protection, export taxes, price guarantees, and income-support programs that encourage the expansion of environmentally unsound practices need to be removed. Much has been accomplished in these areas so far, but there is still more to be done. New national priorities and public expenditure policies will need to incorporate environmental considerations. Impact assessment should become an integral part of national and sectoral policies as well as of project implementation.

With few exceptions, notably the PRC and Viet Nam, which have implemented drastic policy reforms following their open-door policy, past development efforts in Asian countries have emphasized infrastructure development rather than the policy reforms that may improve the effectiveness, transparency, and accountability of government machinery. Technical solutions are preferred to social and economic instruments. The current practice of international lending agencies requiring policy reforms as part of sector loans should be continued. For some countries, this has become the only channel through which sensible policies can be implemented against resistance from groups with vested interest in preserving the status quo. However, the pros and cons of policy reform and the likely impact should be made more transparent to those concerned if not to the general public.

Project Implementation

The reforms described above require action at the policy and institutional levels. Reform is also required at the project implementation level. First, cost-benefit analyses should be rigorously applied during the project inception phase. Second, the potential environmental impact must be fully accounted for and included in the cost-benefit calculation. Third, public participation must be a component of the approval process.

PRIORITY SECTORS

Water and coastal resources are the two natural resource sectors that require immediate action in order to rehabilitate them and to prevent further degradation. Both sectors share similar problems of multiple-use conflicts and overextraction owing to the open-access regimes usually governing their use. Although the crux of these problems is mainly institutional and managerial in nature, investments in rehabilitation and irrigation infrastructure in order to improve irrigation efficiency may be necessary.

For both sectors, development objectives need to be established that are consistent with the goals of long-term ecological balance and sustainability by seeking input from a wide spectrum of stakeholders. The planning and management of these resources should follow the directions given by the third strategy above. Participatory management systems need to be developed that connect the agencies and organizations involved, such as sectoral agencies, local governments, and communities.

Sectoral administrations need to be streamlined in order to eliminate overlapping mandates and jurisdictions. To this end, coordinating mechanisms such as river basin or watershed committees connecting sectoral agencies need to be established.

Coastal areas that are highly crucial for the preservation of biodiversity and ecological balance need to be identified, zoned, and assigned appropriate conservation status. Government agencies in charge of coastal and fisheries resources generally possess expertise that is mostly production oriented. Capacity building in the area of conservation needs to be strengthened.

Appropriate economic instruments, such as transferable quotas, community fishing rights, and concessions should be explored; open-access regimes in coastal and oceanic waters should be replaced. Revised fees for licenses and permits for fishing gear need to reflect the actual economic value derived from the rent of aquatic resources. The number of fishing vessels needs to be regulated to be consistent with sustainable fishing yields, and the use of destructive fishing methods needs to be prohibited. Land-based sedimentation and pollution, as well as the destruction of mangroves, should be controlled and prevented. Strict enforcement of legislation relating to trawling and pushnetting in inshore waters should be implemented.

Research and development of ecologically sound methods of aquaculture, methodologies for fish stock assessment, and the standardization of methodologies and techniques for the domestication of cultivable wild species should all be promoted. Technologies are needed for reducing bycatch, fish discards, and postharvest losses; and landing and primary processing facilities need to be developed.

In the water sector, allocation principles for the use of water in the dry season should be established. For countries where water stress is becoming chronic, demand management will have to be introduced while supply management will have to be strengthened. The polluter-pays principle must be strictly enforced through a combination of economic, legal, and social instruments.

Where the capacity exists to regulate and monitor development and extraction activities, resolve conflicts, and protect coastal and water resources, the appropriate rights and management responsibilities should be devolved to local governments and communities.

Finally, two other priorities beyond the scope of this volume need to be pursued in order to support agricultural sustainability. These are poverty alleviation and human resource development for rural populations. If these two priorities are neglected, it will be difficult to implement successfully the strategies proposed above.

The rapid economic growth of Asia in recent years, especially in the manufacturing and service sectors, has driven some policymakers, both in national governments and international lending agencies, to ignore agriculture and to treat the sector as a sunset industry. Such an attitude in itself is detrimental to the sustainability of agriculture, because reform and rehabilitation require administrative energy and political will. It should also be remembered that since agriculture is very closely related to natural resources and the environment, unsustainable agriculture is often linked to irreversible environmental impact or impact that involves enormous recovery costs.

REFERENCES

Abel, M.E., and S.K. Levin, eds. 1981. *Climate's Impact on Food Supplies: Strategies and Technology for Climate-Defensive Food Production.* Boulder: Westview Press.

ADB (Asian Development Bank). 1995a. *Status of Forestry and Forest Industries in Asia-Pacific Region.* Manila: ADB.

_____.1995b. Fisheries Sector Profile of the People's Republic of China. Agriculture and Social Sectors Department (East), Forestry and Natural Resources Division, Manila: ADB.

_____. 1997a. Emerging Asia: Changes and Challenges. Manila: ADB.

_____.1997b. The Bank's Policy on Fisheries. Manila: ADB.

_____.1998. Agriculture and Environmental Sustainability: Changing Requirements for External Assistance to Agriculture and the Role of the Bank. Agriculture Sector Strategy Review, Ministry of Agriculture, Republic of Indonesia. ADB TA266-IND. Prepared by PT Multi Techniktama Prakarsa and Hunting Technical Services Ltd. LBDS, Jakarta.

Ahmed, M., C. Delgado, and S. Svedrup-Jensen. 1997a. *A Brief for Fisheries Policy Research in Developing Countries.* Manila: International Center for Living Aquatic Resources Management.

_____, A. D. Capistrano, and M. Hossain. 1997b. Experience of Partnership Models for the Co-management of Bangladesh Fisheries. *Fisheries Management and Ecology* 4: 233-248.

Alexandratos, N., ed. 1995. *World Agriculture: Towards 2010. An FAO Study.* Chichester: Wiley & Sons.

Ammar Siamwalla. 1989. Rent Dissipation in Quota Allocations for Cassava in Thailand. Bangkok: Bank of Thailand.

_____.1996. Thai Agriculture: From Engine of Growth to Sunset Status. *TDRI Quarterly Review.* December: 3-10. Bangkok: Thailand Development Research Institute.

_____. 1999. *The Evolving Roles of the Public, Local, and Private Sectors in Rural Asia.* Hong Kong, China: Oxford University Press (China).

_____, Direk Patmasiriwat, Yair Mundlak, and Suthad Setboonsarng. 1989. A Dynamic Analysis of Thai Agricultural Growth: Some Lessons from the Past. Workshop on TDRI Research Activities Supported by EPD II Project, Chiang Mai. 7-8 October. Bangkok: Thailand Development Research Institute.

Anan Kanchanaphan and Mingsarn Kaosa-ard. 1996. Evolution of Settlements in Forest Areas: A Case Study of Upper Northern Thailand. TDRI Research Monograph No. 13. Bangkok: Thailand Development Research Institute. (in Thai)

Anderson, Kym. 1998. Agricultural Trade Reforms, Research Initiatives and the Environment. In *Agriculture and the Environment: Perspectives on Sustainable Rural Development.* Washington DC: World Bank.

APAN (Asia-Pacific Agroforestry Network). 1995. Summary Report of the International Workshop on Agroforestry Investment, Production and Marketing. APAN Report No. 20. Bangkok: APAN.

_____. 1996. Asia-Pacific Agroforestry Profiles: Second Edition. APAN Field Doc. No. 4. RAP Publication No. 1996/20. Bangkok: FAO Regional Office for Asia and the Pacific.

APO (Asian Productivity Organization). 1998. Extension System for Livestock Production in Asia and the Pacific. Report of an APO Multi-Country Study Mission. Tokyo: APO.

AVARD (Association of Voluntary Agencies for Rural Development). 1993. Village-Centred Development in Ralegan Siddhi. New Delhi: AVARD.

Bann, Camille. 1998. The Economic Valuation of Tropical Forest Land Use Options: A Manual for Researchers. Singapore: Economy and Environment Program for South East Asia.

Barker, R., R.W. Herdt, and B. Rose. 1985. *The Rice Economy of Asia.* Washington DC: Resources for the Future.

Beckerman, Wilfred. 1995. A Skeptical View of Sustainable Development. Paper presented at the Conference on Global Agriculture Science Policy for the Twenty First Century, Melbourne, August.

Benge, Michael. D. 1991. *Cambodia: An Environmental and Agricultural Overview and Sustainable Development Strategy.* Washington DC: United States Agency for International Development.

Bhatnagar, P.S. 1995. Soybean Production in India. RAP Publication No. 1995/36. Bangkok: FAO Regional Office for Asia and the Pacific.

Boonlert Phasuk. 1994. Fishing Effort Regulations in the Coastal Fisheries of Thailand. In *Socio-Economic Issues in Coastal Fisheries Management: Proceedings of the IPFC Symposium.* Bangkok: FAO Regional Office for Asia and the Pacific. p. 111-122.

Briggs, M.R.P. 1994. Status, Problems and Solutions for a Sustainable Shrimp Culture Industry. In *Development of Strategies for Sustainable Shrimp Farming, Final Report to the Overseas Development Administration,* edited By M.R.P. Briggs. Research Project R4751. Stirling: Institute of Aquaculture, University of Stirling.

Brown, Lester R. 1995. *Who Will Feed China? Wake up Call for a Small Planet.* New York: Norton & Co.

Bryant, D., Burke, L., McManus, J., and Spalding, M. 1998. Reefs at Risk. Washington DC: World Resources Institute, Manila: International Center for Living Aquatic Resources Management, Cambridge: World Conservation Monitoring Centre, and Nairobi: United Nations Environment Programme.

Bryant, Peter J. 1998. Values of Biodiversity. In *Biodiversity and Conservation: A Hypertext Book.* Available: http://darwin.bio.uci.edu/~sustain/bio65/lec07/ b65lec07.htm.

Bureau of Fisheries, Ministry of Agriculture, PRC. 1997. Protection of Resources and Environment and Fisheries Development in China. In *Environmental Aspects of Responsible Fisheries.* Proceedings of the Asia-Pacific Fishery Commission Symposium. Seoul, 15-18 October 1996. RAPA Publication 1997/32. Bangkok: FAO Regional Office for Asia and the Pacific. p. 52-57.

Byerlee, Derek. 1990. *Technical Change, Productivity and Sustainability in Irrigated Cropping Systems of South Asia: Emerging Issues in the Post-Green Revolution Era.* Economics Working Paper 90/06. Mexico DF: International Center for Maize and Wheat Improvement (CIMMYT).

Cassman, K.G., S. Peng., and A. Dobermann. 1997. Nutritional Physiology of the Rice Plants and Productivity Decline of Irrigated Rice Systems in the Tropics. *Soil Science Plant Nutrition* 43:1101-1106.

CGIAR (Consultative Group on International Agricultural Research). 1998. The International Research Partnership for Food Security and Sustainable Agriculture. Third System Review of the Consultative Group on International Agricultural Research. Washington DC: CGIAR. p. 26-27.

Chambers, Robert, N.C. Saxena, and Tushaar Shar. 1989. *To the Hands of the Poor.* Bombay: Oxford and IBH Publishing Co.

Chand, R., and T. Haque. 1997. Stability of Rice-Wheat Crop System in Indo-Gangetic Region. *Economic and Political Weekly* March 19: A26-A29.

Charoensilp, N., P. Promnart, and P. Charoendham. 1998. An Inter-Regional Programme on Methane Emission from Rice Fields. Paper presented at the Thailand-IRRI Collaborative Research Planning Meeting, Bangkok, 25-26 June.

Chaudhary, M.K., and Harrington, L.W. 1993. The Rice-Wheat System in Haryana: Input-Output Trends and Sources of Future Productivity Growth. Karnal, India: CCS Haryana Agricultural University Regional Research Centre, and Mexico DF: International Center for Maize and Wheat Improvement (CIMMYT).

Chen Zhikai. 1992. Water Resource Development in China. In *Country Experiences with Water resources Management; Economic, Institutional, Technological and Environmental Issues,* edited By Guy Le Moigne et al. Technical paper No.175. Washington DC: World Bank.

_____. 1995. Hybrid Maize in China. RAP Publication No. 1995/ 26. Bangkok: FAO Regional Office for Asia and the Pacific.

Chisholm, Anthony H., and Sisira Jayasuriya. 1994. Economic Growth and Sustainability: Rural China in the Reform Era. Agriculture Economics Discussion Paper 24/94. Agriculture and Resource Economics, School of Agriculture. Melbourne: La Trobe University.

_____, and Sisira Jayasuriya. 1997. Socially Optimal Replacement of Up-country Tea in Sri Lanka. Melbourne: La Trobe University.

_____, Anura Ekanyake, and Jayasuriya Sisira. 1997. Policy and Institutional Reforms and the Environment: The Case of Soil Erosion in Sri Lanka. Sri Lanka Land Degradation Research Project Economic Discussion Paper Series. Colombo.

Coates, D. 1995. Inland Capture Fisheries and Enhancement: Status, Constraints and Prospects for Food Security. Paper presented at the International Conference on the Sustainable Contribution of Fisheries to Food Security, Kyoto. Available: http://www.fao.org/WAICENT/FAOINFO/ FISHERY/ agreem/kyoto/h8f.htm.

Conway, Gordon. 1997. *The Doubly Green Revolution.* London: Penguin.

Coxhead, Ian and Sisira Jayasuriya. 1992. Food Crops, Tree Crops, and Land Degradation in Developing Countries: Effect of Tax on Trade Policies, paper presented at the 48th Congress of International Institute of Public Finance, Seoul, Korea, August 24-27.

_____, and Jiraporn Plangpraphan. 1998. Thailand's Economic Boom and Agricultural Bust: Some Economic Questions and Policy Puzzles. TDRI Quarterly Review: 15-24. Bangkok: Thailand Development Research Institute.

_____, and Agnes Rola. 1998. Economic Development, Agricultural Growth and Environmental Management: What are the Linkages in Lantapan? Paper presented at the SANREM CRSP/Philippines 1998 Annual Conference: *Economic Growth and Natural Resource Management: Are They Compatible?* Bukidnon, Philippines. 18-20 May.

Craswell, E.T. 1998. Sustainable Crop and Soil Management on Sloping Lands. Paper presented at the International Symposium of Asian Agriculture in the 21st Century. Taipei,China: Food and Fertilizer Technology Centre.

Crooks, R. 1995. Water Quality and the Environment in Water Resources Sector Review. A joint report by ADB, FAO, UNDP, Water Resources Group in cooperation with the Institute of Water Resources Planning.

Crosson, Pierre. 1994. Degradation of Resources as a Threat to Sustainable Agriculture. Paper prepared for the First World Congress of Professionals in Agronomy, Santiago, Chile, 5-8 September.

_____. 1995. Soil Erosion and Its On-Farm Productivity Consequences: What Do We Know? Discussion Paper 95-29. Washington DC: Resources for the Future.

Csavas, I. 1992. Impacts of Aquaculture on the Shrimp Industry. In *Proceedings of the 3rd Global Conference on the Shrimp Industry, Shrimp '92*, edited by H. deSaram and T. Singh. Hong Kong: Infofish. p. 6-18.

David, Cistina C., and Keijiro Otsuka. 1994. *Modern Rice Technology and Income Distribution in Asia*. Boulder: Lynne Rienner,and Los Baños, Philippines: International Rice Research Institute.

de Foresta, H. and Michon, G. 1994. From Shifting Cultivation to Forest Management through Agroforestry: Smallholder Damar Agroforests in West Lampung (Sumatra). *APANews* 6/7: 12-16.

_____. 1997. The Agroforest Alternative to *Imperata* Grasslands: When Smallhholder Agriculture and Forestry Reach Sustainability. *Agroforestry Systems* 36:105-120.

Deb Menasveta. 1997. Fisheries Management Frameworks of the Countries Bordering the South China Sea. RAP Publication 1997/33. Bangkok: FAO Regional Office for Asia and the Pacific.

Desai, G.M. and V. Gandhi. 1989. Phosphorus for Sustainable Agricultural Growth in Asia: An Assessment of Alternative Sources and Management. Symposium on Phosphorus Requirements for Sustainable Agriculture in Asia and Pacific Region, International Rice Research Institute, Los Baños, Philippines, 6-10 March.

Devaraj, M., and E. Vivekanandan. 1997. A Comparative Account of the Small Pelagic Fisheries in the APFIC Region. In *Small Pelagic Resources and Their Fisheries in the Asia-Pacific Region*, edited By M. Devaraj and P. Martosubroto. Proceedings of the APFIC Working Party on Marine Fisheries, First Session, 13-16 May 1997, Bangkok, Thailand. RAP Publication 1997/31. Bangkok: FAO Regional Office for Asia and the Pacific. p. 17-61

Dierberg, F. E., and Woraphan Kiattisimkul. 1996. Issues, Impacts, and Implications of Shrimp Aquaculture in Thailand. *Environmental Management* 20(5): 649-666.

Direk Patmasiriwat, O. Kuik and S. Pednekar. 1998. The Shrimp Aquaculture Sector in Thailand: A Review of Economic, Environmental and Trade Issues. CREED Working Paper

No. 19. London: International Institute for Environment and Development, and Amsterdam: Institute for Environmental Studies, Free University.

Economy and Environment Program for Southeast Asia (EEPSEA) and the World Wide Fund for Nature (WWF). 1998. New Estimates Place Damage from Indonesia's 1997 Forest Fires at $4.4 Billion. Press Release. May.

Engelman, Robert, and Pamela LeRoy. 1995. *Conserving Land: Population and Sustainable Food Production*. Population and Environment Program. Washington DC: Population Action International.

Evenson, Robert. 1996. Valuing Agricultural Biodiversity. In *Global Agricultural Science Policy for the Twenty First Century*. Conference Proceedings. Melbourne: Department of Natural Resources and Environment. p. 611-638.

Fairbairn, Donald K. 1995. Did the Green Revolution Concentrate Incomes? A Quantitative Study of Research Reports. *World Development* 23(2): 265-279.

Falkenmark, Malin, and Carl Widstrand. 1992. Population and Water Resources: A Delicate Balance. *Population Bulletin* 47(3): 2-34.

—————, Jan Lundqvist, and Carl Widstrand. 1989. Macro-Scale Water Scarcity Requires Micro-Scale Approaches: Aspects of Vulnerability in Semi Arid Development. National Resource Forum. London: Butterworth. p. 258-267.

Fan, Shenggen. 1997. Production and Productivity Growth in Chinese Agriculture: New Measurement and Evidence. *Food Policy* 22: 213-228.

—————, Peter Hazeil, and Sukhadeo Thorat. 1998. Government Spending, Growth and Poverty: An Analysis of Interlinkages in Rural India. EPTD Discussion Paper No. 33. Environment and Production Technology Division, Washington DC:International Food Policy Research Institute.

Fay, Chip, Hubert de Foresta , Martua Sirat, and Thomas P.Tomich. 1998. A Policy Breakthrough for Indonesian Farmers in the Krui Damar Agroforests. *Agroforestry Today* 10: 25-26.

FAO (Food and Agriculture Organization of the United Nations). 1995a. Dimensions of Need: An Atlas of Food and Agriculture. Rome: FAO.

_____. 1995b. Forest Resources Assessment 1990: Global Synthesis. Rome: FAO.

_____.1997a. State of the World's Forests 1997. Rome: FAO.

_____.1997b. Review of the State of World Fishery Resources. FAO Fisheries Circular No. 920. Rome: FAO.

_____. 1997c. The State of World Fisheries and Aquaculture 1996. Rome: FAO.

_____.1998a. Proceeding of the Regional Workshop on Area-Wide Integration of Crop-Livestock Activities. RAP Publication 1998/19. Bangkok: FAO Regional Office for Asia and the Pacific.

_____.1998b. FISHSTAT PC Software. Data Series 1950-1996. Fishery Information, Data and Statistics Unit, Fisheries Department. Rome: FAO.

_____.1998c. AQUACULT PC Software. Data Series 1984-1996. Fishery Information, Data and Statistics Unit, Fisheries Department. Rome: FAO.

_____.1998d. Number of Fishers Doubled Since 1970. Available: http://www.fao.org/WAICENT/FAOINFO/FISHERY/highligh/fisher/fisher.htm. Accessed: December 1998. Rome: FAO.

_____. 1998e. Aquaculture Institute a Catalyst for Blue Revolution in India. Available: http://www.fao.org/news/1998/980802%2De.htm. Accessed Nov. 1998. Rome: FAO.

Flinn, J.C., S.K. de Datta, and E. Labadan. 1980. An Analysis of Long-term Rice Yields in a Wetland Soil. *Field Crop Research* 5: 201-216.

Frisvold, G.B., and P.T. Condon. 1998. The Convention on Biological Diversity and Agriculture: Implications and Unresolved Debates. *World Development* 26: 515-570.

Fry, James. 1998. The Competitiveness of Sugar Production. Paper presented at the 4th annual Asia International Sugar Conference, Phuket, Thailand, 9-11 Sept.

Fullen, M.A., and D.J. Mitchell. 1994. Desertification and Reclamation in North Central China. *Ambio* XXIII(2): 131-135.

Garrity, D.P. 1998. Participatory Approaches to Catchment Management: Some Experiences to Build Upon. Paper presented at the Managing Soil Erosion Consortium Assembly, Hanoi, Viet Nam, June 8-12.

Garrity, D.P., et al. 1997. The *Imperata* Grasslands of Tropical Asia: Area, Distribution, and Topology. *Agroforestry Systems* 36:3-29.

_____, and Patricio C. Agustin. 1995. Historical Land Use Evolution in a Tropical Acid Upland Agroecosystem. *Agriculture, Ecosystem and Environment* 53: 83-95.

Ghani, Md.A., S.I. Bhuiyan, and R.W. Hill. 1989. Gravity Irrigation Management in Bangladesh. *Journal of Irrigation Drainage Engineering* 115: 642-655.

Gill, Gerard J. 1995. Major Natural Resource Management Concerns in South Asia. Food, Agriculture and the Environment Discussion Paper No. 8. Washington DC: International Food Policy Research Institute.

Goldman, Abe, and Joyatee Smith. 1995. Agricultural Transformations in India and Northern Nigeria: Exploring the Nature of Green Revolutions. *World Development* 23(2): 243-263.

Gopakumar, K. 1997. Environmental Impact of Harvesting Techniques and Utilization of Discards in India. In Environmental Aspects of Responsible Fisheries. Proceedings of the Asia-Pacific Fishery Commission Symposium. Seoul, 15-18 October 1996. RAPA Publication 1997/32. Bangkok: FAO Regional Office for Asia and the Pacific. p. 58-85.

GRAIN. 1997. Engineering the Blue Revolution. *Seedling* 14 (4): 20-30.

Grainger R. J. R., and S. M. Garcia. 1996. Chronicles of Marine Landings (1950-1994): Trend Analysis and Fisheries Potential. FAO Fisheries Technical Paper No. 359. Rome: Food and Agriculture Organization of the United Nations.

Guerra, L.C., S.I.Bhuiyan, T.P.Tuong, and R. Barker. 1998. Producing More Rice With Less Water From Irrigated Systems. Available: www.cgiar.org/swim.

de Haan, Cees, Henning Steinfeld, and Harvey Blackburn. 1997. Livestock and the Environment. Finding a Balance. Brussels: European Commission, Directorate-General for Development, Development Policy, Sustainable Development and Natural Resources.

Handa, B.K. 1983. Effects of Fertiliser Use on Groundwater Quality in India. In *Groundwater Resources Planning* II. IAHS publication No. 142. p. 1105-1109.

Hardin, Garrett. 1968. The Tragedy of the Commons. *Science* 162: 1243-1248.

Harrington, Larry. 1983. Sustainable Agriculture for Uplands in Asia: Direct vs. Preventative Contributions of New Technology. Paper presented at the CGPRT Centre Regional Seminar on Upland Agriculture in Asia, Bogor, Indonesia,6-8 April.

Hartwick, J.M. 1977. Intergenerational Equity and the Investing of Rents from Exhaustible Resources. *American Economic Review* 66: 972-974.

Hawkes, J. 1985. Plant Genetic Resources: The Impact of the International Agricultural Research Centers. CGIAR Study Paper No 3. Washington DC: World Bank.

Hazell, Peter, and Shenggen Fan. 1998. Balancing Regional Development Priorities to Achieve Sustainable and Equitable Agricultural Growth. Paper prepared for the AAEA International Conference on Agricultural Intensification, Economic Development and the Environment, Salt Lake City, Utah, 31 July - 1 August.

_____, and C. Ramaswamy. 1991. *The Green Revolution Reconsidered: The Impact of High Yielding Rice Varieties in South India.* Baltimore: Johns Hopkins University Press.

Hobbs, P. R., and M.L. Morris. 1996. Meeting South Asia's Further Food Requirements From Rice-Wheat Cropping Systems: Priority Issues Facing Researchers in the Post Green Revolution Era. Natural Resource Group Paper 96-01. Mexico DF: International Center for Maize and Wheat Improvement (CIMMYT).

Hobbs, P.R., K.D. Sayre, and J.I. Ortiz-Monsterio. 1998. Increasing Wheat Yields Sustainably through Agronomic Means. Natural Resource Group Paper No. 98-01. Mexico DF: International Center for Maize and Wheat Improvement (CIMMYT).

Hong, Yang. 1996. Trends in China Regional Grain Production and Their Implications. Working Paper No. 96/10. Chinese Economies Research Centre. Adelaide: University of Adelaide.

Hossain, M. 1996. Economic Prosperity in Asia: Implications for Rice Research. In Proceedings of the Third International Rice Genetics Symposium. Los Baños, Philippines: International Rice Research Institute.

_____, and P. L. Pingali. 1998. Rice Research, Technological Progress and Impact on Productivity and Poverty: an Overview. In *Impact of Rice Research*, edited By P.L. Pingali and M. Hossain. Bangkok: Thailand Development Research Institute, and Los Baños, Philippines: International Rice Research Institute. p. 1-25

Hotta, M. 1996. Regional Review of the Fisheries and Aquaculture Situation and Outlook in South and Southeast Asia. FAO Fisheries Circular No. 904. Rome: Food and Agriculture Organization of the United Nations.

ICLARM (International Center for Living Aquatic Resources Management). 1998. Dissemination and Evaluation of Genetically Improved Tilapia Species in Asia: Final Report. Manila: ICLARM.

_____. 1999. ICLARM's Strategic Plan 2000-2020. Manila: ICLARM.

ICRAF (International Centre for Research in Agroforestry). 1996. Annual Report. Nairobi: ICRAF.

International Commission of Irrigation and Drainage. 1991. Country Report on Irrigation and Drainage Development in Pakistan.

IPCC (Intergovernmental Panel on Climate Change). 1996. Climate Change 1995. Impacts, Adaptations and Mitigation of Climate Change: Scientific-Technical Analyses. Contribution of Working Group II to the Second Assessment Report of the Intergovernmental Panel on Climate Change.

IRG/Edgevale/REECS. 1996. The Philippine Environmental and Natural Resources Accounting Project, Phase III. Main report. Prepared for the Department of Environment and Natural Resources and the United States Agency for International Development. Manila: Philippine Environment and Natural Resources Accounting Project.

IRRI (International Rice Research Institute). 1983. International Rice Research Institute Annual Report for 1982. Los Baños, Philippines: IRRI.

Jantakad, P., and S. Carson. 1998. Community Based Natural Resource Management from Villages to an Inter-Village Network: a Case study in Pang Ma Pha District, Mae Hong Son Province, Northern Thailand. Thai-German Highland Development Project.

Johnson, Gale D. 1996. China's Rural and Agricultural Reforms: Successes and Failures. Working Paper No. 96/12. Chinese Economies Research Centre. Adelaide: University of Adelaide.

Kanok Rerkasem, Benjavan Rekasem, Mingsarn Kaosa-ard, Chaiwat Roonruangsee, S. Jesdapipat, S. Shinawatra, and P. Wijukprasert. 1994. Sustainability of Highland Agriculture. Bangkok: Thailand Development Research Institute.

_____,Benjavan Rerkasem, Mingsarn Kaosa-ard, Chaiwat Roonruangsee, and Adis Israngkura. 1989. Highland Development as a Narcotic Prevention Strategy. Report prepared for the U.S. Agency for International Development, Bangkok. Chiang Mai: Chiang Mai University.

Kerr, John, Ganesh Pangare, Vasudha Lokur Pangare, P.J. George Kolavalli, and Shashi Kolavalli. 1998. The Role of Watershed Projects in Developing Rainred Agriculture in India. Prepared for the Indian Council for Agricultural Research and The World Bank.

_____,Peter Hazell, and Dayanatha Jha. 1998. Sustainable Development of Rainfed Agriculture in India. A Synthesis of Findings of the World Bank-ICAR Project on Sustainable Rainfed Agriculture Research and Development. Prepared for the Indian Council of Agricultural Research. Washington DC: International Food Policy Research Institute.

Khan, A.Z.M. Obaidullah. 1996. Asian Agriculture and the New Millennium. In House Discussion Paper. Bangkok: FAO Regional office for Asia and the Pacific.

Kikuchi, M. 1996. Present and Future of Irrigation Development in Developing Countries in Asia. Paper presented at the Workshop held at Hokkaido University, 15-16 September.

Kumar, Praduman, and Mark W. Rosegrant. 1994. Productivity and Sources of Growth for Rice in India. *Economic and Political Weekly* 31 December: A183-188.

Lan Yisheng, and Peng Zaoyang. 1997. China's Fishery Industry: Production, Consumption and Trade. Working Paper No. 97/9. Chinese Economies Research Centre. Adelaide: University of Adelaide.

Lin, Yifu Justin. 1992. Rural Reforms and Agriculture Growth in China. *American Economic Review* 82(1): 34-51.

_____. 1998a. How Did China Feed Itself in the Past and How Could China Achieve It Again in the Future? Peking University, and Hong Kong, China: Hong Kong University of Science and Technology.

————. 1998b. Technological Change and Agricultural and Resource Economics. *Australian Journal of Agriculture and Resource Economics.* Forthcoming.

————, and Minggao Shen. 1994. Rice Production Constraints in China: Implications for Biotechnology Initiative. Workshop on Rice Research Prioritization in Asia, Los Baños, Philippines, 21-22 February. New York: Rockefeller Foundation, and Los Baños, Philippines: International Rice Research Institute.

Lindert, Peter H. 1996a. Soil Degradation and Agricultural Change in two developing countries. In *Global Agricultural Science Policy for the Twenty First Century.* Conference Proceedings. Melbourne: Department of Natural Resources and Environment. p. 263-332.

————. 1996b. The Bad Earth? China's Agricultural Soils Since the 1930's. Working Paper Series No. 83. Agricultural History Center. Davis: University of California.

Litvak, J. 1995. Water Resources and Poverty in Vietnam. In *Water Resources Sector Review.* Manila: Asian Development Bank, Rome: FAO, New York: United Nations Development Programme, and Hanoi: Institute of Water Resources Planning.

Loevinsohn, M.E. 1987. Insecticide Use and Increased Mortality in Rural Central Luzon, Philippines. *Lancet* 13 June: 1359-1362.

Longworth, J.W. and G.J. Williamson. 1993. *China's Pastoral Region.* London: CAB International, and Canberra: Australian Centre for International Agricultural Research.

Lohani, Bindu M. 1998. Environmental Challenges in Asia in the 21st Century. Environment Division. Manila: Asian Development Bank (Mimeo).

Lu, Z.G., R.W. Bell, L. Huang, D. Hu, and Z.C. Xie. 1997. Nutrient Status of Rape and Soils in Main Rape Production Areas of Hubei Province. *Hubei Agricultural Sciences* 3: 28-32.

Magliano, A.R., and A.R. Librero, 1998. Indigenous Technical Knowledge on Soil Management and Sustainable Agriculture in the Philippines. Issues in Sustainable Land Management No. 3: 63-83. Bangkok: International Board on Soil Research and Management.

Magrath W.B., and P. Arens. 1987. *The Cost of Soil Erosion on Java. A Natural Resource Accounting Approach.* Washington DC: World Resources Institute.

Malik, R.K., and S. Singh. 1994. Effects of Biotypes and Environment on the Efficacy of Isoproturon Against Wild Canary Grass. *Annals of Applied Biology* 124 (supplement). Test of Agrochemicals and Cultivars 15: 52-53.

_____, G. Gill, and P.R. Hobbs. 1998. Herbicide Resistance - a Major Issue for Sustaining Wheat Productivity in Rice-Wheat Cropping Systems in the Indo-Gangetic Plains. Rice-Wheat Consortium Paper Series 3. Mexico DF: International Center for Maize and Wheat Improvement (CIMMYT), and Los Baños, Philippines: International Rice Research Institute.

Manwan, I., K. Suradisastra, D.A. Adjid, and R. Montgomery. 1998. The Implication of Technological Change. Agriculture Sector Strategic Review. Jakarta: Ministry of Agriculture, Republic of Indonesia. ADB TA2660-INO. PT Multi Thniktama Prakarsa, Hunting Technical Services Ltd., Lembaga Bangun Desa Sejatera.

Matthews, R.B. et al. 1994a. Climate Change and Rice Production in Asia. *Entwicklung und Landlicherraum* 1: 16-19.

Matthews, R.B. et al. 1994b. The Impact of Global Climate Change on Rice Production in Asia: a Simulation Study. Report No. ERL-COR-821. Environmental Research Laboratory. Corvallis: U.S. Environmental Protection Agency.

Mao, C. 1994. Hybrid Rice Production in China: Status and Strategies. In *Hybrid Research and Development Needs in Major Cereals in the Asia Pacific Region,* edited By R.S. Paroda and M. Rai. RAP Publication No. 1994/21. Bangkok: FAO Regional Office for Asia and the Pacific. p. 87-93.

McGrath, W.B., and P. Arens. 1987. *The Cost of Soil Erosion on Java - a Natural Resource Accounting Approach.* Washington DC: World Resources Institute.

McNeely, Jeffrey, A. 1998. Mobilizing Broader Support for Asia's Biodiversity: How Civil Society Can Contribute to Protected Area Management. Gland: IUCN-The World Conservation Union.

_____, Keaton R. Miller, Walter V. Reid, Russel A. Mittermeier, and Timothy B. Werner. 1990. *Conserving the World's Biological Diversity.* Washington DC: World Bank and World

Resources Institute, and Gland: World Conservation Union (IUCN), Conservation International, and World Wildlife Fund.

———, J. Harrison, and P. Dingwall, eds. 1994. *Protecting Nature:Regional Reviews of Protected Areas.* Gland: International Conservation Union (IUCN).

Meyer, Richard L., and Geetha Nagarajan. 1999. *Rural Financial Markets in Asia: Policies, Paradigms, and Performance.* Hong Kong, China: Oxford University Press (China).

Mie Xie. 1996. Water Resources in Vietnam. In *Water Resources Sector Review.* Manila: Asian Development Bank, Rome: FAO, New York: United Nations Development Programme, and Hanoi: Institute of Water Resources Planning.

Mingsarn Kaosa-ard. 1995. Rules, Instruments and Public Participation for Environmental Protection. Thailand Development Research Institute (TDRI), Year End Conference, Jomtien, Chonburi, 9-10 December. Bangkok: TDRI.

———. 1997. Sharing the Benefits and Cost of Forest Conservation. In *Environment and Development in the Pacific: Problems and Policy Options,* edited by H. Edward English and David Runnalls. Melbourne: Addison Wesley Longman. p. 73-85.

———, and Ammar Siamwalla. 1997. Formulation of the Chao Phraya Basin Water Resources Management Strategy. Institutional and Legal Framework for Water Resource Management. Unpublished report. Bangkok: Thailand Development Research Institute (in Thai).

———, and S. Pednekar. 1998. Background Report for the Thai Marine Rehabilitation Plan 1997-2001. Bangkok: Thailand Dement Research Institute.

———, Kanok Rerkasem, and Chaiwat Roonruangsee. 1989. Agricultural Information and Technological Change in Northern Thailand. TDRI Research Monograph No. 1. Bangkok: Thailand Development Research Institute.

——— et al. in collaboration with Theodore Panayotou and J.R. Deshazo. 1995. Green Finance: A Case Study of Khao Yai. Bangkok: Thailand Development Research Institute.

——— et al. 1999. A Decade of Change: Natural Resources and Environment of Thailand. Bangkok: Thailand Development Research Institute. Forthcoming.

_____, Kobkun Rayanakorn, Gerard Cheong, Suzaune White, Craig A. Johnson, and Pinida Kongsiri. 1998. Towards Public Participation in the Mekong Basin Development Plan. Report prepared for the Mekong River Commission. Bangkok: Thailand Development Research Institute.

Moench, Marcus. 1994. Approach to Ground Water Management: To Control or Enable? *Economic and Political Weekly* 24 September: A135-A145.

Morris, R.A. 1997. Vegetables, Nutrient Rates and Nutrient Management. In *Managing Soil Fertility for Intensive Vegetable Production Systems in Asia,* edited By R.A. Morris. AVRDC Pub. No. 97-469. Taipei,China: Asian Vegetable Research and Development Center. p. 5-24.

Morris, M.L., R.P. Singh, and S. Pal. 1998. Indian Maize Seed Industry in Transition: Changing Roles for the Public and Private Sector. *Food Policy* 23(1).

MRC (Mekong River Commission).1997. Mekong River Basin Diagnostic Study. Final Report MKG/R.97010. Bangkok: MRC.

Murdiyarso, Daniel. 1998. *Impact* 2(2): 1-4.

Murray-Rust, D.H., and W.B. Snellen. 1993. *Irrigation System Performance Assessment and Diagnosis.* Colombo: International Irrigation Management Institute.

Nangju, Dimyati. 1996. Potential Constraints and Prospects for Vegetable Production in Asia. Speech delivered at the Workshop on Collaborative Vegetable Research in Southeast Asia. Proceedings of the AVNET II Final Workshop, Bangkok,1-6 September.

NRC (National Research Council). 1989. *Alternative Agriculture.* Washington DC: National Academy Press.

_____. 1991. *Managing Global Genetic Resources: the US National Plant Germplasm System.* Washington DC: National Academy Press.

Oldeman, L.R. 1992. Global Extent of Soil Degradation. Biannual Report. Wageningen: International Soil Reference and Information Centre.

Oldeman, L.R., R.T.A. Hakkeling, and W.G. Sombroek. 1990. World Map of the Status of Human-induced Soil Degradation: An Explanatory Note. Nairobi: International Soil Reference and Information Centre, and United Nations Environment Programme.

Oka , I.N. 1996. Status of Integrated Pest Management: Progress and Problems. APO Study Meeting Report on Intergrated Pest Management in Asia and the Pacific. Tokyo: Asian Productivity Organization.

Paine J.R., N. Bryron, and M. Poffenberger. 1997. Status, Trends and Future Scenarios for Forest Conservation including Protected Areas in the Asia-Pacific Region. Working Paper No. APFSOS/WP/04. Forestry Policy and Planning Division. Rome: FAO.

Panayotou, Theodore. 1993. *Green Markets: The Economics of Sustainable Development.* San Francisco: ICS Press.

Paroda, R.S., and K.L. Chadha, eds. 1996. 50 Years of Crop Science Research in India. New Delhi: Indian Council of Agricultural Research.

Pauly, Daniel, Villy Christensen, Peter Dalsgaard, Rainer Froese, and F. Torres Jr. 1998. Fishing Down Marine Food Webs. *Science* 279: 860-863.

Pimentel, D. et al. 1995. Environmental and Economic Costs of Soil Erosion and Conservation Benefits. *Science* 261: 1117-1123.

Pingali, P.L., and S. Rajaram. 1998. Technological Opportunities for Sustaining Wheat Productivity Growth Toward 2020. IFPRI 2020 Brief 51. Washington DC: International Food Policy Research Institute.

—————, and Mark W. Rosegrant. 1993. Confronting the Environmental Consequences of the Green Revolution in Asia. Paper presented at the 1993 AAEA International Pre-conference on Post-Green Revolution Agricultural Development Strategies in the Third World: What Next? August.

Pongsroypech, C. 1994. Hybrid Maize and Sorghum Development in Thailand. In *Hybrid Research and Development Needs in Major Cereals in the Asia Pacific Region,* edited by R.S. Paroda, and M. Rai. RAP Publication No. 1994/21. Bangkok: FAO Regional Office for Asia and the Pacific. p. 193-219.

Pray, Carl E. 1991. The Development of Asian Research Institutions: Underinvestment, Allocation of Resources, and Productivity. In *Research and Productivity in Asian Agriculture,* edited by Robert E. Evenson and Carl E. Pray. Ithaca: Cornell University Press. p. 47-80.

Qureshi, R.H., and E.G. Barrett-Lennard. 1998. Saline Agriculture for Irrigated Land in Pakistan: A Handbook. ACIAR Monograph 50. Canberra: Australian Centre for International Agricultural Research.

Rafiq, M. 1990. Soil Resources and Soil Related Problems in Pakistan. In *Soil Physics - Application Under Stress Environments*, edited by M. Ahmed. Islamabad: BARD, PARC.

Ramsay, G.C, and L. Andrews. 1999. The Role of Livestock in Food Security in the Region. RAP Publication. Bangkok: FAO Regional Office for Asia and the Pacific. (in press)

Ratana, Chuenpagdee. 1998. Damage Schedules for Thai Coastal Areas: An Alternative Approach to Assessing Environmental Values. Economy and Environment Program for Southeast Asia (EEPSEA) Research Report Series. Singapore: EEPSEA.

Rattan, Vernon W., ed. 1994. *Agriculture, Environment and Health: Sustainable Development in the 21st Century*. Minneapolis: University of Minnesota Press.

Reid, Walter V.,Ana Sitenfeld, Sarah A. Laird, Daniel H. Janzen, Carrie A. Meyer, Michael A. Gollin, Rodrigo. Gamez, and Calestous Juma.1993. *Biodiversity Prospecting: Using Genetic Resources for Sustainable Development*. Washington DC: World Resources Institute, Costa Rica: Institut Nacional de Biodiversidad (INBio),USA: Rainforest Alliance, Kenya: Africa Centre for Technology Studies.

Repetto, Robert. 1994. *The 'Second India' Revisited: Population, Poverty and Environmental Stress Over Two Decades*. Washington DC: World Resources Institute.

Reynolds, Stephen G. 1995. Pasture, Cattle, Coconut Systems. RAP Publication 1995/7. Bangkok: FAO Regional Office for Asia and the Pacific.

Rola, A., and P.L. Pingali. 1993. Pesticides, Health Risks and Farm Productivity: A Philippines Experience. Los Baños, Philippines: International Rice Research Institute.

Rosegrant, Mark W., and Peter B.R. Hazell. 1999. *Transforming the Rural Asian Economy: the Unfinished Revolution*. Hong Kong China: Oxford University Press (China).

————, and P.L. Pingali. 1991. Sustaining Rice Productivity Growth in Asia: A Policy Perspective. IRRI Social Science Division Paper No. 91-01. Los Baños, Philippines: International Rice Research Institute.

_____, and Claudia Ringler. 1997. World Food Markets into the 21st Century: Environmental and Resource Constraints and Policies. *Australian Journal of Agriculture and Resource Economics* 41(3): 401-428.

_____, and Claudia Ringler. 1998. Impact on Food Security and Rural Development of Reallocating Water from Agriculture. Paper presented at the Harare Expert Group Meeting on Strategic Approaches to Freshwater Management, Zimbabwe, 28-31January.

Royal Government of Cambodia. 1998. Cambodia: National Environmental Action Plan 1998-2002. Phnom Penh.

Rozelle, S., J. Huang, and L. Zhang. 1997. Poverty, Population and Environmental Degradation in China. *Food Policy* 22: 229-252.

Ruckes, E. 1996. Future Availability of Fishery Products for Human Consumption in Asia. In Report of the Regional Consultation on Institutional Credit for Sustainable Fish Marketing, Capture and Management in Asia and the Pacific, Manila, Philippines, 3-7 July. FAO Fisheries Report No. 540. Rome: Food and Agriculture Organization of the United Nations. p. 51-65.

Saleth, R. Maria, and Ariel Dinar. 1999. *Water Challenge and Institutional Response: A Cross-Country Perspective.* Washington DC: World Bank.

Sam, D.D. 1994. Shifting Cultivation in Vietnam. Country Report, IIED Forestry and Land Use Series No. 3. 65p. London: International Institute for Environment and Development.

Seckler, David, Upali Amarasinghe, David Molden, Radhika de Silva, and Randolph Barker. 1998. World Water Demand and Supply, 1990 to 2025: Scenarios and Issues. Research Report 19. Colombo: International Water Management Institute.

Sen, Amartya. 1986. Food Economics and Entitlements. World Institute for Development Economics Research. Beirut: United Nations University.

Serrano, R.C. 1988. The Sloping Agricultural Land Technology: Case analysis of a development program. *IESAM Bulletin* 8(3): 10-14.

Shah, P.B. 1997. Technology Development in Rainfed Agriculture. In Report of an APO Study on Rainfed Agriculture in Asia. Tokyo: Asian Productivity Organization. p. 71-88.

Sharma, Rita, and Thomas H. Poleman. 1993. *The New Economics of India's Green Revolution. Income and Employment Diffusion in Uttar Pradesh.* Ithaca: Cornell University Press.

Silvestre, Geronimo, and Daniel Pauly. 1997. Management of Tropical Coastal Fisheries in Asia: An Overview of Key Challenges and Opportunities. In *Status and Management of Tropical Coastal Fisheries in Asia,* edited by Geronimo Silvestre and Daniel Pauly. ICLARM Conference Proceedings 53. Manila: International Center for Living Aquatic Resources Management. p. 8-25.

Singh, I.P., and Derek Byerlee. 1990. Relative Variability in Wheat Yields Across Countries and Over Time. *Journal of Agricultural Economics* 41: 21-32.

State Ministry of Environment, Republic of Indonesia, and KONPHALINDO. 1995. *An Atlas of Biodiversity in Indonesia.* Jakarta.

Steinfield, Henning, Cornelis de Haan, and Harvey Blackburn. 1998. Livestock and the Environment: Issues and Options. In *Agriculture and the Environment: Perspectives on Sustainable Rural Development.* Washington: World Bank.

Stone, B. 1986. Chinese Fertilizer Application in the 1980's and 1990's: Issues of Growth, Balances, Allocation, Efficiency and Response. In *China's Economy Looks Toward the Year 2000.* Volume 1. Washington DC: Congress of the United States. p. 453-493.

Suraphol, Sudara. 1997. Marine Fisheries and Environment in the ASEAN Region, In Environmental Aspects of Responsible Fisheries. Proceedings of the Asia-Pacific Fishery Commission (APFIC) Symposium, Seoul, 15-18 October 1996. RAPA Publication 1997/32. Bangkok: FAO Regional Office for Asia and the Pacific. p. 184-205.

TAC (Technical Advisory Committee of the CGIAR). 1997. Priorities and Strategies for Soil and Water Aspects of Natural Resources Management Research in the CGIAR. TAC Secretariat. Rome: Food and Agriculture Organization of the United Nations.

Tacio, H.D. 1990. A Sloping Agriculture Land Technology (SALT): a Sustainable Agroforestry Scheme for Uplands. *Agroforestry Systems* 22: 145-152.

Tanaka, A., and S. Yoshida. 1970. Nutritional Disorders of the Rice Plants in Asia. IRRI Technical Bulletin No. 10. Los Baños, Philippines: International Rice Research Institute.

Thapan Kumar Mishra. 1998. Users Become Managers: Indigenous Knowledge and Foreign Forestry. *Economic and Political Weekly* 7 February: 262-3.

Tarumizu, Kimimasa. 1992. President's Opening Address. In Proceedings of the Regional Workshop on Sustainable Agricultural Development in Asia and the Pacific Region, Manila and Los Baños, Philippines, June 15-19. Edited by Denise Felton Bryant.

TDRI (Thailand Development Research Institute). 1997. Review of the Suitability and the Feasibility of the Kaeng Sua Ten Dam, Phrae Province. Report submitted to the Budget Bureau, The Royal Thai Government. Bangkok: TDRI.

————, and HIID (Harvard Institute for International Development). 1995. Green Finance: A Case Study of Khao Yai National Park. Natural Resource and Environment Program. Bangkok: Thailand Development Research Institute, and Cambridge: HIID, Harvard University.

Ticsay-Ruscoe, M. 1995. The Use of Formal Questionnaires and Interview Schedules for Biodiversity Study: the Case of Barangays Haliap-Panubtuban, Aspolo, Ifugao, Central Cordillera, Northern Luzon, Philippines. In *Biodiversity: Concepts, Frameworks and Methods. Proceedings of a Regional Study*, edited by P. Shenji, and P.E. Sajise. Kunming: Yunan University Press. p. 41-52.

Tomich, Thomas, P., A.M. Fagi, H. de Foresta, G. Michon, D. Murdiyarso, F. Stolle, and Meine van Noordwijk. 1998a. Indonesia's Fires: Smoke as a Problem, Smoke as a Symptom. *Agroforestry Today* 10(1): 4-7.

————, Meine van Noordwijk, Stephen A. Vosti, and Julie Witcover, 1998b. Agricultural Development with Rainforest Conservation: Methods for Seeking Best Bet Alternatives to Slash and Burn, with Applications to Brazil and Indonesia. *Agricultural Economics* 19:159-174.

Toribio, M.Z.B., and J.L. Orno. 1995. The Indigenous Agroforestry System Production of the Ifugao. In IDCJ Research Report. Tokyo: International Development Center of Japan. p. 68-79.

Turner II, B.L., and P. Benjamin. 1994. Fragile Lands: Identification and Use for Agriculture. Vernon W. Rattan (ed.) In *Agriculture, Environment and Health: Sustainable Development in the 21st Century*, edited by Vernon W. Rattan. Minneapolis: University of Minnesota Press. p. 104-143.

UNEP (United Nations Environment Programme). 1997. *Global Environment Outlook*. New York: Oxford University Press.

US Embassy, Yangon. 1996. Foreign Economic Trends Report: Burma. Yangon.

Viswanathan, K. K., N. M. R. Abdullah, I. Susilowati, I. M. Siason, and C. Ticao. 1997. Enforcement and Compliance with Fisheries Regulations in Malaysia, Indonesia and the Philippines. Fisheries Co-Management Project Research Report No. 5. Manila: International Center for Living Aquatic Resources Management, and Denmark: The North Sea Centre.

Vivekanandan, P. 1998. New Tree, New System: Neem Production in South India. *Agroforestry Today* 10(1): 12-14.

Vermillion, Douglas L. 1997. Impact of Irrigation Management Transfer: A Review of the Evidence. Research Report No. 11. Colombo: International Irrigation Management Institute.

Vyas, V.S., and V. Ratna Reddy. 1998. Assessment of Environmental Policies and Policy Implementation in India. *Economic and Political Weekly*. 10 January, p. 48-54.

Waddington, S.R., J.K. Ransom, M. Osmanzai, and D.A. Saunders. 1986. Improvement in the Yield Potential of Bread Wheat Adapted to Northwest Mexico. *Crop Science* 26: 698-703.

Wang, L., and Y. Guo. 1994. Rice-Wheat Systems and Their Development in China. In *Sustainability of Rice-Wheat Production Systems in Asia*, edited by R.S. Paroda, T. Woodhead, and R.B. Singh. Bangkok: FAO Regional Office for Asia and the Pacific. p. 160-171.

Wang, Y. 1996. Supply, Demand and Sustainable Growth of Grains in China. Paper presented at the Science Academic Summit. Uncommon Opportunities for a Food Secure World, Madras, 8-11 July.

Wassman, R. T.B. Moya, T.B. and R.S. Lantin, R.S. 1998. Rice and the Global Environment. *Field Crops Research* 56.

Watson, H. 1987. Sloping Agricultural Land Technology: a Case Study of Upland Farming in Southern Philippines. Lecture presented at UPLB, Philippines, 4 November. Los Baños: University of the Philippines at Los Baños.

Wei, Y.Z., R.W. Bell, Y. Yang, J.M. Xue, K. Wang, and L. Huang. 1998. Prognosis of Boron Deficiency in Oilseed Rape (*Brassica napus L.*) by Plant Analysis. *Australia Journal of Agricultural Research* 49: 867-874.

Weirsum, K.F. 1984. Surface Erosion Under Various Tropical Agroforestry Systems. In Proceedings of the Symposium on the Effects of Forest Land Use on Erosion and Slope Stability. Honolulu: East West Center. p.231-239.

Williams, M. 1996. The Transition in the Contribution of Living Aquatic Resources to Food Security. Washington DC: International Food Policy Research Institute.

_____. and M.A. Bimbao. 1998. Aquaculture: The Last Frontier for Sustainable Food Security? Dean D.K. Villaluz Memorial Lecture Tigbauan, Philippines: Southeast Asian Fisheries Development Center.

World Bank. 1996. Indonesia: Decentralized Agricultural Extension Study. Paper prepared by P.T. Mitra Lingkungan Dinamika, Jakarta.

WCED (World Commission on Environment and Development). 1987. *Our Common Future*. Oxford: Oxford University Press.

WRI (World Resources Institute). 1994. World Resources 1994/95. Washington DC: WRI.

_____. 1997. World Resources 1996-97. New York: Oxford University Press.

_____. 1998. World Resources 1998/99. Washington DC: WRI.

Yamamoto, T. 1998. Community-based Fisheries Management. In *Community-based Fisheries Management in Phang-nga Bay, Thailand*, edited by D.J. Nickerson. RAP Publication 1998/3. Bangkok: FAO Regional Office for Asia and the Pacific. p. 209-227.

Yuan, L.P. 1994. Increasing Yield Potential in Rice by Exploitation of Heterosis. In *Hybrid Rice Technology: New Developments and Future Prospects*, edited by S.S. Virmani. Los Baños, Philippines: International Rice Research Institute. p. 1-6.

Yin Runsheng. 1997. Forestry and the Environment in China. In *Environment and Development in the Pacific : Problems and Policy Options*, edited by H. Edward English and David Runnalls. Melbourne: Addison Wesley Longman. p. 52-72.

van Zalinge, Nicolaas, Nao Thuok, Touch Seang Tana, and Diep Loeung. 1998. Where There is Water, There is Fish?

Cambodian Fisheries Issues in a Mekong River Basin Perspective. Project for the Management of the Freshwater Capture Fisheries of Cambodia. Bangkok: Mekong River Commission, Phnom Penh: Cambodia Department of Fisheries, and Copenhagen: Danish International Development Agency.

Zhang, Qishun, M., and Zhang Xiao. 1995. Water Issues and Sustainable Social Development in China. *Water International* 20(3): 122-128.

Annex A

Tables on Population and Agricultural and Fisheries Production in Asia

Table A1: Population Growth in Asia

	Population ('000 persons)[a]			Average Growth (percent per year)[b]	
	1977	1987	1997	1977-1986	1987-1997
World	4,227,160	5,020,682	5,848,731	1.72	1.55
Asia	2,447,900	2,951,890	3,538,452	1.87	1.82
East Asia	**1,110,316**	**1,270,006**	**1,417,661**	**1.35**	**1.13**
China, People's Rep. of	958,438	1,104,216	1,243,738	1.41	1.23
Japan	113,882	122,078	125,638	0.75	0.31
Korea, Rep. of	36,465	41,681	45,717	1.39	0.93
Mongolia	1,531	2,031	2,568	2.79	2.42
Southeast Asia	**337,521**	**416,487**	**496,843**	**2.12**	**1.79**
Cambodia	6,824	7,919	10,516	0.94	2.87
Indonesia	141,748	173,666	203,480	2.06	1.61
Lao PDR	3,097	3,819	5,194	1.89	3.08
Malaysia	12,831	16,541	21,018	2.50	2.42
Myanmar	31,776	39,073	46,765	2.08	1.81
Philippines	45,081	57,091	70,724	2.38	2.14
Thailand	43,587	52,996	59,159	2.03	1.16
Viet Nam	50,259	62,550	76,548	2.20	2.03
South Asia	**851,677**	**1,062,962**	**1,291,153**	**2.21**	**1.97**
Afghanistan	15,892	14,165	22,132	-0.90	3.97
Bangladesh	81,143	103,670	122,013	2.53	1.67
Bhutan	1,209	1,531	1,862	2.30	2.03
India	647,230	801,1	93960,178	2.13	1.84
Maldives	145	196	273	2.93	3.34
Nepal	13,447	17,370	22,591	2.55	2.62
Pakistan	78,539	108,350	143,831	3.13	2.88
Sri Lanka	14,072	16,487	18,273	1.63	1.05
Central Asia			**55,250**		**1.18**
Kazakhstan			16,832		-0.01
Kyrgyz Republic			4,481		0.15
Uzbekistan			23,656		1.98
Tajikistan			6,046		1.78
Turkmenistan			4,235		1.98

[a] Population is the average in that year.
[b] Average annual growth rate is calculated by (ln (end year) - ln (begin year)) * 100/ number of years in period.

Growth rates of Central Asian countries represent 1992-1997.

Source: FAOSTAT Agriculture Data. Population. 28 August 1998. *Available: http://apps.fao.org*

Table A2: Cereal and Pulse Production in Asia, 1977-1997

Crop	Production[a]		Average Growth (percent per year)	
	1977	1997	1977-1986	1987-1997
Rice				
Million t	330.305	509.445	3.48	1.99
Million ha	127.205	132.673	0.11	0.40
t/ha	2.596	3.840	3.35	1.52
Wheat				
Million t	91.204	198.582	7.86	3.69
Million ha	59.209	66.844	1.19	0.55
t/ha	1.540	2.970	6.22	2.97
Maize				
Million t	69.088	147.440	4.31	4.98
Million ha	34.620	40.406	0.23	1.03
t/ha	1.994	3.647	4.04	3.44
Pulses				
Million t	21.540	24.108	0.67	1.43
Million ha	34.057	35.460	-0.03	1.31
t/ha	0.633	0.680	0.72	0.07
Sorghum				
Million t	19.970	15.765	-0.194	1.43
Million ha	20.577	13.222	-0.398	-1.93
t/ha	0.971	1.191	0.013	1.36
Millet				
Million t	16.705	14.088	-0.139	-0.83
Million ha	23.751	15.887	-0.440	-1.85
t/ha	0.703	0.886	0.010	1.48
Barley				
Million t	8.362	6.235	-0.102	-1.22
Million ha	5.151	3.061	-0.117	-2.27
t/ha	1.623	2.035	0.030	1.83

[a]   three-year running average

Source: FAOSTAT Database. Available: http://apps.fao.org

Table A3: Asia's Other Crops[a], 1977–1997

Crop	Production		Average Growth (percent per year)	
	1977[b]	1996	1977-1986	1987-1997
Oils (excluding oil palm)				
Million t	51.389	106.400	4.28	4.68
Million ha	39.849	62.417	2.96	2.29
t/ha	1.290	1.705	1.05	1.99
Soybean				
Million t	8.557	20.958	6.56	5.33
Million ha	8.364	15.239	3.48	3.76
t/ha	1.022	1.376	2.40	1.23
Peanuts				
Million t	9.587	20.769	4.75	4.19
Million ha	10.466	13.415	1.39	0.82
t/ha	0.916	1.549	2.97	3.10
Cottonseed				
Million t	7.789	18.140	9.15	2.89
Rapeseed				
Million t	3.600	16.276	15.67	8.27
Million ha	6.431	13.984	4.70	4.72
t/ha	0.559	1.164	7.68	2.40
Sunflower				
Million t	0.330	2.937	66.71	5.47
Million ha	0.464	3.177	44.15	3.72
t/ha	0.706	0.925	5.11	1.36
Sesame				
Million t	1.013	1.717	5.97	1.82
Million ha	3.848	4.090	1.93	-0.39
t/ha	0.263	0.420	3.56	2.33
Linseed				
Million t	0.629	0.845	-0.93	-1.05
Million ha	2.222	1.701	-2.61	2.10
t/ha	0.282	0.497	2.22	-3.22
Castor				
Million t	0.363	1.084	9.99	7.39
Million ha	0.705	1.072	4.62	1.47
t/ha	0.518	1.012	4.14	5.00
Safflower				
Million t	0.215	0.471	12.64	2.14
Million ha	0.688	0.738	3.43	-2.39
t/ha	0.314	0.642	7.31	6.22
Vegetables				
Million t	124.342	335.268	6.27	6.84
Million ha	10.941	20.814	3.76	4.01
t/ha	11.361	16.109	1.92	1.96

(continued next page)

Table A3 (continued)

Crop	Production		Average Growth (percent per year)	
	1977[b]	1996	1977-1986	1987-1997
Beverages				
Million t	1.760	3.679	4.99	3.74
Million ha	2.654	3.971	2.38	2.37
t/ha	0.663	0.926	2.27	1.03
Tobacco				
Million t	2.261	4.383	5.71	3.26
Million ha	1.944	2.671	2.45	1.30
t/ha	1.161	1.633	2.85	1.58
Nuts				
Million t	0.654	1.123	2.84	3.76
Million ha	0.580	1.192	7.25	2.77
t/ha	1.127	0.941	-2.72	0.69
Cotton				
Million t	3.913	9.156	9.10	1.09
Million ha	11.688	27.265	9.14	2.93
Roots and tubers				
Million t	215.731	255.941	0.15	1.95
Million ha	17.460	16.358	-1.21	0.53
t/ha	12.347	15.643	1.53	1.31
Potato				
Million t	40.070	75.862	2.05	6.46
Million ha	3.187	5.217	2.30	3.36
t/ha	12.567	14.537	-0.14	2.34
Sweet potato				
Million t	133.403	128.707	-1.29	1.13
Million ha	10.518	7.185	-3.21	-0.34
t/ha	12.678	17.913	2.68	1.51
Cassava				
Million t	39.669	48.211	2.92	-0.68
Million ha	3.468	3.645	1.42	-0.77
t/ha	11.427	13.232	1.47	0.14

[a]   includes fibers, oils, roots, sugar, coffee, tea, tobacco, rubber, vegetables, fruits, and nuts
[b]   three-year running average

Source: FAOSTAT database. *Available: http://apps.fao.org*

Table A4: Diversification in Asia's Cropping Systems, 1977-1997

	Cropping System	Area (ha million)		Area (percent of total)	
		1977[a]	1997	1977[a]	1997
East Asia					
China, People's Rep. of	Foodgrains	101.286	93.516	69.6	53.0
	Others	44.283	82.780	30.4	46.9
Japan	Foodgrains	2.982	2.245	75.4	75.7
	Others	0.975	0.720	24.6	24.3
Korea, Rep. of	Foodgrains	1.380	1.104	68.0	72.0
	Others	0.649	0.430	32.0	28.0
Southeast Asia					
Cambodia	Foodgrains	0.988	2.013	91.3	91.5
	Others	0.094	0.188	8.7	8.5
Indonesia	Foodgrains	11.052	15.554	63.2	53.9
	Others	6.440	13.327	36.8	46.1
Lao PDR	Foodgrains	0.624	0.606	92.6	83.0
	Others	0.050	0.124	7.4	17.0
Malaysia	Foodgrains	0.748	0.696	20.6	10.5
	Others	2.878	5.911	79.4	89.5
Myanmar	Foodgrains	5.849	8.068	76.3	76.2
	Others	1.813	2.519	23.7	23.8
Philippines	Foodgrains	6.896	6.658	60.1	55.7
	Others	4.582	5.296	39.9	44.3
Thailand	Foodgrains	9.526	11.046	76.6	70.0
	Others	2.910	4.742	23.4	30.0
Viet Nam	Foodgrains	5.737	7.850	81.5	75.8
	Others	1.302	2.512	18.5	24.2
South Asia					
Afghanistan	Foodgrains	3.117	2.084	88.3	88.7
	Others	0.412	0.267	11.7	11.3
Bangladesh	Foodgrains	11.378	11.502	87.3	85.1
	Others	1.651	2.016	12.7	14.9
Bhutan	Foodgrains	0.094	0.092	86.1	79.6
	Others	0.015	0.024	13.9	20.4
India	Foodgrains	122.59	125.47	67.2	58.9
	Others	59.739	87.426	32.8	41.1
Nepal	Foodgrains	2.388	3.541	89.5	87.4
	Others	0.280	0.511	10.5	12.6
Pakistan	Foodgrains	11.381	13.887	78.1	64.6
	Others	3.1956	7.597	21.9	35.4
Sri Lanka	Foodgrains	0.790	0.827	38.9	45.4
	Others	1.240	0.995	61.1	54.6

Source: FAOSTAT Database. *Available: http://apps.fao.org*

Table A5: Contributions to Total Crop Area From Individual Crops, 1977 and 1997

Crop	Percent of Total Crop Area[a]		Percent Change
	1977[b]	1997	1977-1997
Rice	30.84	27.44	-3.40
Wheat	14.35	13.82	-0.53
Maize	8.39	8.36	-0.04
Sorghum	1.25	0.63	-0.62
Millet	4.99	2.73	-2.25
Pulses	8.26	7.33	-0.92
Cereals + Pulses	73.84	63.60	-10.24
Oils[c]	9.89	14.20	4.31
Fibers	3.72	6.19	2.47
Roots	4.23	3.38	-0.85
Sugar	1.52	1.90	0.38
Beverages	0.64	0.82	0.18
Tobacco	0.47	0.55	0.08
Rubber	1.24	1.34	0.11
Vegetables	2.65	4.30	1.65
Fruits	1.66	3.46	1.81
Nuts	0.14	0.25	0.11
Others	26.16	36.40	10.24

[a]   harvested area
[b]   three-year moving average
[c]   including oil palm

Source: FAOSTAT Database. *Available: http://apps.fao.org*

Table A6: Rice Yield and Yield Growth in 1977 and 1997, by Country

	Yield (t/ha)		Average Growth (percent per year)	
	1977[a]	1997	1977-1986	1987-1997
East Asia				
China, People's Rep. of	3.704	6.187	5.22	1.77
Japan	5.948	6.433	0.41	0.58
Korea, Rep. of	6.564	6.545	0.53	0.49
Southeast Asia				
Cambodia	1.033	1.474	2.68	0.63
Indonesia	2.821	4.477	5.00	1.05
Lao PDR	1.249	2.541	7.18	2.40
Malaysia	2.639	3.032	0.00	1.61
Myanmar	1.981	3.204	6.92	0.99
Philippines	1.945	2.892	4.45	0.91
Thailand	1.797	2.219	1.63	0.98
Viet Nam	1.988	3.689	4.18	3.39
South Asia				
Afghanistan	2.025	1.765	1.14	-0.99
Bangladesh	1.887	2.636	1.91	1.61
Bhutan	2.000	1.589	-0.27	-0.69
India	1.863	2.871	2.41	2.44
Nepal	1.850	2.284	0.21	0.32
Pakistan	2.367	2.780	0.75	2.05
Sri Lanka	2.122	3.365	5.44	1.33
Asia	**2.596**	**3.840**	**3.35**	**1.50**
	1988	1992		
USA	6.175	6.413		
Australia	6.975	8.813		
		1990-1995		
Punjab		3.353		
Haryana		2.759		

[a] three-year running average

Source: FAOSTAT Database. *Available: http://apps.fao.org*

Table A7: Wheat Yield and Yield Growth in 1977 and 1997, by Country

	Yield (t/ha)		Average Growth (percent per year)	
	1977[a]	1997	1977-1986	1987-1997
East Asia				
China, People's Rep. of	1.695	3.787	8.73	3.33
Japan	2.840	3.196	2.61	-0.35
Korea, Rep. of	1.999	4.134	6.38	3.75
Southeast Asia				
Myanmar	0.852	1.088	11.96	-2.55
South Asia				
Afghanistan	1.193	1.146	0.15	-0.30
Bangladesh	1.639	1.962	4.97	1.35
Bhutan	1.004	0.738	0.18	0.37
India	1.426	2.533	4.26	3.10
Nepal	1.112	1.555	2.69	1.75
Pakistan	1.390	2.051	2.60	2.37
Asia	**1.562**	**3.003**	**6.21**	**2.96**
		1990-95	·	
Punjab	3.878			
Haryana		3.599		

[a]   three-year running average

Source: FAOSTAT Database. *Available: http://apps.fao.org*

Table A8: Maize Yield and Yield Growth in 1977 and 1997, by Country

	Yield (t/ha)		Average Growth (percent per year)	
	1977[a]	1997	1977-1986	1987-1997
East Asia				
China, People's Rep. of	2.609	4.867	5.07	2.76
Japan	2.722	2.476	-2.81	0.12
Korea, Rep. of	2.768	4.148	8.98	-1.64
Southeast Asia				
Cambodia	1.295	1.259	-1.97	3.19
Indonesia	1.261	2.394	5.11	2.45
Lao PDR	1.170	2.328	-0.15	7.76
Malaysia	1.458	1.816	1.41	0.26
Myanmar	0.882	1.569	11.99	-0.20
Philippines	0.895	1.568	2.87	4.46
Thailand	2.076	3.219	1.41	4.41
Viet Nam	1.085	2.384	3.48	7.20
South Asia				
Afghanistan	1.620	1.812	0.39	1.00
Bangladesh	0.846	1.026	-0.09	0.75
Bhutan	1.400	0.874	-0.25	0.44
India	1.062	1.589	2.09	2.76
Nepal	1.697	1.668	-0.69	0.98
Pakistan	1.235	1.437	0.41	0.72
Sri Lanka	0.699	1.004	6.29	-1.28
Asia	**1.994**	**3.647**	**4.04**	**3.34**
	1988	1992		
USA	5.313	8.250		
France	7.169	7.888		

[a]   three-year running average

Source: FAOSTAT Database. *Available: http://apps.fao.org*

Table A9: Harvested Area of Coconuts in Asia, 1977-1997

	Harvested Area (ha)			Average Growth (percent per year)	
	1977	1987	1997	1978-1987	1988-1997
East Asia	0	13,000	29,320		8.13
China, People's Rep. of	0	13,000	29,320		8.13
Southeast Asia	5,006,200	6,235,039	6,776,274	2.20	0.83
Brunei	238	90	90	-9.72	0.00
Cambodia	0	8,000	11,000		3.18
Indonesia	1,655,000	2,096,762	2,547,761	2.37	1.95
Malaysia	344,000	320,591	260,000	-0.70	-2.09
Myanmar	17,806	26,738	31,033	4.07	1.49
Philippines	2,713,960	3,252,000	3,314,390	1.81	0.19
Thailand	223,200	331,558	352,000	3.96	0.60
Viet Nam	51,996	199,300	260,000	13.44	2.66
South Asia	1,535,839	1,795,145	2,285,755	1.56	2.42
Bangladesh	26,867	31,560	32,036	1.61	0.15
India	1,056,500	1,346,000	1,810,000	2.42	2.96
Maldives	1,000	1,000	1,300	0.00	2.62
Pakistan	0	162	560		12.40
Sri Lanka	451,472	416,423	441,859	-0.81	0.59
Total Asia	6,544,043	8,043,934	9,091,364	2.06	1.22

Source: FAOSTAT Database. Agricultural Production Indexes. 22 April 1999. *Available: http://apps.fao.org*

Table A10: Output of Coconuts in Asia, 1977-1997

	Output (t)			Average Growth (percent per year)	
	1977	1987	1997	1978-1987	1988-1997
East Asia	56,000	80,000	116,970	3.57	3.80
China, People's Rep. of	56,000	80,000	116,970	3.57	3.80
Southeast Asia	20,808,930	24,835,079	30,827,220	1.77	2.16
Brunei	285	140	130	-7.11	-0.74
Cambodia	42,000	42,000	58,000	0.00	3.23
Indonesia	8,150,000	1,920,000	14,710,000	2.93	2.98
Malaysia	1,123,000	1,022,000	967,000	-0.94	-0.55
Myanmar	92,668	229,500	209,300	9.07	-0.92
Philippines	10,281,000	10,520,000	12,052,790	0.23	1.36
Thailand	927,400	1,310,486	1,430,000	3.46	0.87
Viet Nam	192,577	790,953	1,400,000	14.13	5.71
South Asia	5,484,580	7,236,491	11,903,655	2.77	4.98
Bangladesh	68,497	82,615	89,255	1.87	0.77
India	4,022,000	5,402,000	9,800,000	2.95	5.96
Maldives	10,083	9,560	13,000	-0.53	3.07
Pakistan	0	316	2,400		20.27
Sri Lanka	1,384,000	1,742,000	1,999,000	2.30	1.38
Total Asia	26,361,910	32,156,650	42,847,971	1.99	2.87

Source: FAOSTAT Database. Agricultural Production Indexes. 22 April 1999. *Available: http://apps.fao.org*

Table A11: Yield of Coconuts in Asia, 1977-1997

	Yield (kg/ha)			Average Growth (percent per year)	
	1977	1987	1997	1978-1987	1988-1997
East Asia					
China, People's Rep. of	0	6,154	3,989		-4.33
Southeast Asia	**3,748**	**4,992**	**5,148**	**2.86**	**0.31**
Brunei	1,198	1,556	1,444	2.62	-0.74
Cambodia	0	5,250	5,273		0.04
Indonesia	4,925	5,208	5,774	0.56	1.03
Malaysia	3,265	3,188	3,719	-0.24	1.54
Myanmar	3,204	8,583	6,744	5.00	-2.41
Philippines	3,788	3,235	3,637	-1.58	1.17
Thailand	4,155	3,953	4,063	-0.50	0.27
Viet Nam	3,704	3,969	5,385	0.69	3.05
South Asia	**4,876**	**5,581**	**6,753**	**1.35**	**1.91**
Bangladesh	2,550	2,618	2,786	0.26	0.62
India	3,807	4,103	5,414	0.53	2.99
Maldives	10,083	9,560	10,000	-0.53	0.45
Pakistan	0	1,951	4,286	-0.53	0.45
Sri Lanka	3,066	4,183	4,524	3.11	0.78
Asia	**4,028**	**3,998**	**4,713**	**-0.08**	**1.65**

Source: FAOSTAT Database. Agricultural Production Indexes. 22 April 1999. *Available: http://apps.fao.org*

Table A12: Harvested Area of Rubber in Asia, 1977-1997

	Harvested Area (ha)			Average Growth (percent per year)	
	1977	1987	1997	1978-1987	1988-1997
East Asia	**0**	**365,000**	**395,000**		**0.79**
China, People's Rep. of	0	365,000	395,000		0.79
Southeast Asia	**4,658,835**	**5,140,648**	**5,774,123**	**0.98**	**1.16**
Brunei	3,936	3,200	2,800	-2.07	-1.34
Cambodia	20,000	30,000	45,000	4.05	4.05
Indonesia	1,557,000	1,884,033	2,260,471	1.91	1.82
Malaysia	1,800,000	1,535,000	1,470,000	-1.59	-0.43
Myanmar	46,539	40,677	45,852	-1.35	1.20
Philippines	58,540	84,038	92,879	3.62	1.00
Thailand	1,094,720	1,360,000	1,527,721	2.17	1.16
Viet Nam	78,100	203,700	329,400	9.59	4.81
South Asia	**412,563**	**442,200**	**559,200**	**0.69**	**2.35**
Bangladesh	0	0	27,000		
India	186,000	237,100	374,000	2.43	4.56
Sri Lanka	226,563	205,100	158,200	-1.00	-2.60
Total Asia	**5,071,398**	**5,947,848**	**6,728,323**	**1.59**	**1.23**

Source: FAOSTAT Database. Agricultural Production Indexes. 22 April 1999. *Available: http://apps.fao.org*

Table A13: Output of Rubber in Asia, 1977-1997

	Output (t)			Average Growth (percent per year)	
	1977	1987	1997	1978-1987	1988-1997
East Asia	91,700	238,000	451,970	9.54	6.41
China, People's Rep. of	91,700	238,000	451,970	9.54	6.41
Southeast Asia	3,004,417	4,015,029	5,243,550	2.90	2.67
Cambodia	15,000	25,000	40,000	5.11	4.70
Indonesia	853,978	1,130,351	1,548,609	2.80	3.15
Malaysia	1,588,053	1,578,700	1,082,400	-0.06	-3.77
Myanmar	14,940	15,031	25,961	0.06	5.46
Philippines	58,198	147,247	197,160	9.28	2.92
Thailand	430,900	1,067,000	2,168,720	9.07	7.09
Viet Nam	43,348	51,700	180,700	1.76	12.51
South Asia	295,843	341,306	650,783	1.43	6.45
Bangladesh	0	0	3,000		
India	149,600	219,500	542,000	3.83	9.04
Sri Lanka	146,243	121,806	105,783	-1.83	-1.41
Total Asia	3,392,165	4,594,520	6,346,483	3.03	3.23

Source: FAOSTAT Database. Agricultural Production Indexes. 22 April 1999. *Available: http://apps.fao.org*

Table A14: Yield of Rubber in Asia, 1977-1997

	Yield (kg/ha)			Average Growth (percent per year)	
	1977	1987	1997	1978-1987	1988-1997
East Asia	0	652	1,144		5.62
China, People's Rep. of	0	652	1,144		5.62
Southeast Asia	606	775	951	2.46	2.05
Brunei	406	578	643	3.53	1.07
Cambodia	750	833	889	1.05	0.65
Indonesia	549	600	685	0.89	1.32
Malaysia	882	1,029	736	1.54	-3.35
Myanmar	321	370	566	1.42	4.25
Philippines	994	1,752	2,123	5.67	1.92
Thailand	394	785	1,420	6.89	5.93
Viet Nam	555	254	549	-7.82	7.71
South Asia	752	760	743	0.47	-0.23
Bangladesh	0	0	111		
India	804	926	1,449	1.41	4.48
Sri Lanka	646	594	669	-0.84	1.19
Asia	6,689	7,725	9,432	1.44	2.00

Source: FAOSTAT Database. Agricultural Production Indexes. 22 April 1999. *Available: http://apps.fao.org*

Table A15: Harvested Area of Tea in Asia, 1977-1997

	Harvested Area (ha)			Average Growth (percent per year)	
	1977	1987	1997	1978-1987	1988-1997
East Asia	**1,101,752**	**893,113**	**912,700**	**-2.10**	**0.22**
China, People's Rep. of	1,041,832	832,663	860,000	-2.24	0.32
Japan	59,700	59,900	51,800	0.03	-1.45
Korea, Rep. of	220	550	900	9.16	4.92
Southeast Asia	**179,125**	**213,948**	**266,871**	**1.78**	**2.21**
Indonesia	79,079	98,463	114,287	2.19	1.49
Lao PDR	70	190	395	9.99	7.32
Malaysia	2,734	3,355	2,953	2.05	-1.28
Myanmar	51,000	55,400	61,236	0.83	1.00
Thailand	3,200	13,000	17,000	14.02	2.68
Viet Nam	43,042	43,540	71,000	0.12	4.89
South Asia	**654,454**	**679,783**	**679,553**	**0.38**	**0.00**
Bangladesh	42,998	45,785	48,308	0.63	0.54
India	369,184	411,700	440,000	1.09	0.66
Nepal	260	800	772	11.24	-0.36
Sri Lanka	242,012	221,498	190,473	-0.89	-1.51
Total Asia	**2,013,684**	**1,899,558**	**2,021,793**	**-0.58**	**0.62**

Source: FAOSTAT Database. Agricultural Production Indexes. 22 April 1999. *Available: http://apps.fao.org*

Table A16: Output of Tea in Asia, 1977-1997

	Output (t)			Average Growth (percent per year)	
	1977	1987	1997	1978-1987	1988-1997
East Asia	**380,904**	**631,378**	**728,697**	**5.05**	**1.43**
China, People's Rep. of	278,403	534,578	636,497	6.52	1.75
Japan	102,301	96,300	91,200	-0.60	-0.54
Korea, Rep. of	200	500	1,000	9.16	6.93
Southeast Asia	**121,034**	**177,918**	**230,673**	**3.85**	**2.60**
Indonesia	82,928	126,096	149,463	4.19	1.70
Lao PDR	65	502	110	20.44	-15.18
Malaysia	4,542	4,830	6,000	0.61	2.17
Myanmar	14,800	14,100	17,700	-0.48	2.27
Thailand	400	3,400	5,100	21.40	4.05
Viet Nam	18,299	28,990	52,300	4.60	5.90
South Asia	**802,274**	**873,012**	**1,143,077**	**0.84**	**2.70**
Bangladesh	37,022	37,595	53,310	0.15	3.49
India	556,267	620,800	810,000	1.10	2.66
Nepal	413	1,290	2,906	11.39	8.12
Sri Lanka	208,572	213,327	276,861	0.23	2.61
Total Asia	**1,406,904**	**1,866,014**	**2,344,204**	**2.82**	**2.28**

Source: FAOSTAT Database. Agricultural Production Indexes. 22 April 1999. *Available: http://apps.fao.org*

Table A17: Yield of Tea in Asia, 1977-1997

	Yield (kg/ha)			Average Growth (percent per year)	
	1977	1987	1997	1978-1987	1988-1997
East Asia	**2,890**	**3,159**	**3,612**		
China, People's Rep. of	267	642	740	8.77	1.42
Japan	1,714	1,608	1,761	-0.64	0.91
Korea, Rep. of	909	909	1,111	0.00	2.01
Southeast Asia	**686**	**1,091**	**824**	**4.64**	**-2.80**
Indonesia	1,049	1,281	1,308	2.00	0.21
Lao PDR	929	2,642	279	10.45	-22.48
Malaysia	1,296	1,440	2,032	1.05	3.44
Myanmar	290	255	289	-1.29	1.25
Thailand	125	262	300	7.40	1.35
Viet Nam	425	666	737	4.49	1.01
South Asia	**1,205**	**1,226**	**2,041**	**0.18**	**5.09**
Bangladesh	861	821	1,104	-0.48	2.96
India	1,507	1,508	1,841	0.01	2.00
Nepal	1,589	1,613	3,764	0.15	8.47
Sri Lanka	862	963	1,454	1.11	4.12
Asia	**6,987**	**9,823**	**11,595**	**3.41**	**1.66**

Source: FAOSTAT Database. Agricultural Production Indexes. 22 April 1999. *Available: http://apps.fao.org*

Table A18: Harvested Area of Green Coffee in Asia, 1977-1997

	Harvested Area (ha)			Average Growth (percent per year)	
	1977	1987	1997	1978-1987	1988-1997
East Asia	**12,071**	**20,000**	**23,000**	**5.05**	**1.40**
China, People's Rep. of	12,071	20,000	23,000	5.05	1.40
Southeast Asia	**495,116**	**888,374**	**1,279,009**	**5.85**	**3.64**
Cambodia	280	205	350	-3.12	5.35
Indonesia	380,327	652,518	831,782	5.40	2.43
Lao PDR	6,605	14,242	23,345	7.68	4.94
Malaysia	13,450	15,875	14,000	1.66	-1.26
Myanmar	2,478	3,002	4,110	1.92	3.14
Philippines	76,200	143,232	149,455	6.31	0.43
Thailand	8,726	37,920	65,967	14.69	5.54
Viet Nam	7,050	21,380	190,000	11.09	21.85
South Asia	**166,921**	**257,230**	**258,460**	**4.32**	**0.05**
India	160,000	243,500	242,000	4.20	-0.06
Sri Lanka	6,921	13,730	16,460	6.85	1.81
Total Asia	**682,665**	**1,183,574**	**1,592,087**	**5.50**	**2.97**

Source: FAOSTAT Database. Agricultural Production Indexes. 22 April 1999. *Available: http://apps.fao.org*

Table A19: Output of Green Coffee in Asia, 1977-1997

	Output (t)			Average Growth (percent per year)	
	1977	1987	1997	1978-1987	1988-1997
East Asia	8,052	26,000	48,000	11.72	6.13
China, People's Rep. of	8,052	26,000	48,000	11.72	6.13
Southeast Asia	320,506	599,938	1,087,080	6.27	5.94
Cambodia	85	130	280	4.25	7.67
Indonesia	193,966	388,669	453,956	6.95	1.55
Lao PDR	4,315	5,312	12,300	2.08	8.40
Malaysia	5,400	11,500	10,000	7.56	-1.40
Myanmar	940	1,488	1,696	4.59	1.31
Philippines	105,100	140,119	130,000	2.88	-0.75
Thailand	6,300	25,220	78,548	13.87	11.36
Viet Nam	4,400	27,500	400,300	18.33	26.78
South Asia	112,687	198,130	216,348	5.64	0.88
India	102,300	192,100	205,000	6.30	0.65
Sri Lanka	10,387	6,030	11,348	-5.44	6.32
Total Asia	446,155	829,179	1,361,753	6.20	4.96

Source: FAOSTAT Database. Agricultural Production Indexes. 22 April 1999. *Available: http://apps.fao.org*

Table A20: Yield of Green Coffee in Asia, 1977-1997

	Yield (kg/ha)			Average Growth (percent per year)	
	1977	1987	1997	1978-1987	1988-1997
East Asia	667	1,300	2,087	6.67	4.73
China, People's Rep. of	667	1,300	2,087	6.67	4.73
Southeast Asia	662	719	896	1.46	2.20
Cambodia	304	634	800	7.35	2.33
Indonesia	510	596	546	1.56	-0.88
Lao PDR	653	373	527	-5.60	3.46
Malaysia	402	724	714	5.88	-0.14
Myanmar	379	496	413	2.69	-1.83
Philippines	1,379	978	870	-3.44	-1.17
Thailand	722	665	1,191	-0.82	5.83
Viet Nam	624	1,286	2,107	7.23	4.94
South Asia	1,070	614	768	-5.55	2.24
India	639	789	847	2.11	0.71
Sri Lanka	1,501	439	689	-12.29	4.51
Asia	654	701	855	0.69	1.99

Source: FAOSTAT Database. Agricultural Production Indexes. 22 April 1999. *Available: http://apps.fao.org*

Table A21: Harvested Area of Oil Palm in Asia, 1977-1997

	Harvested Area (ha)			Average Growth (percent per year)	
	1977	1987	1997	1978-1987	1988-1997
East Asia	0	**50,500**	**42,000**		**-1.84**
China, People's Rep. of	0	50,500	42,000		-1.84
Southeast Asia	**671,246**	**1,878,905**	**4,123,503**	**10.29**	**7.86**
Indonesia	140,000	421,600	1,622,503	11.02	13.48
Malaysia	521,486	1,373,147	2,317,000	9.68	5.23
Philippines	6,000	15,000	19,000	9.16	2.36
Thailand	3,760	69,158	165,000	29.12	8.70
Total Asia	**671,246**	**1,929,405**	**4,165,503**	**10.56**	**7.70**

Source: FAOSTAT Database. Agricultural Production Indexes. 22 April 1999. *Available: http://apps.fao.org*

Table A22: Output of Oil Palm in Asia, 1977-1997

	Output (t)			Average Growth (percent per year)	
	1977	1987	1997	1978-1987	1988-1997
East Asia	**480,000**	**670,000**	**608,000**	**3.33**	**-0.97**
China, People's Rep. of	480,000	670,000	608,000	3.33	-0.97
Southeast Asia	**9,987,900**	**32,526,462**	**73,137,000**	**11.81**	**8.10**
Indonesia	2,400,000	8,859,147	26,800,000	13.06	11.07
Malaysia	7,500,000	22,800,000	43,700,000	11.12	6.51
Philippines	42,000	139,000	237,000	11.97	5.34
Thailand	45,900	728,315	2,400,000	27.64	11.92
Total Asia	**10,467,900**	**33,196,462**	**73,745,000**	**11.54**	**7.98**

Source: FAOSTAT Database. Agricultural Production Indexes. 22 April 1999. *Available: http://apps.fao.org*

Table A23: Yield of Oil Palm in Asia, 1977-1997

	Yield (kg/ha)			Average Growth (percent per year)	
	1977	1987	1997	1978-1987	1988-1997
East Asia	0	**13,267**	**14,476**		**0.87**
China, People's Rep. of	0	13,267	14,476		0.87
Southeast Asia	**12,683**	**14,354**	**15,600**	**1.24**	**0.83**
Indonesia	17,143	21,013	16,518	2.04	-2.41
Malaysia	14,382	16,604	18,861	1.44	1.27
Philippines	7,000	9,267	12,474	2.81	2.97
Thailand	12,207	10,531	14,546	-1.48	3.23
Asia	**15,595**	**17,206**	**17,704**	**0.98**	**0.29**

Source: FAOSTAT Database. Agricultural Production Indexes. 22 April 1999. *Available: http://apps.fao.org*

Table A24: Harvested Area of Sugar Cane in Asia, 1977-1997

	Harvested Area (ha)			Average Growth (percent per year)	
	1977	1987	1997	1978-1987	1988-1997
East Asia	662,472	960,164	1,125,700	3.71	1.59
China, People's Rep. of	629,872	925,264	1,103,200	3.85	1.76
Japan	32,600	34,900	22,500	0.68	-4.39
Southeast Asia	1,333,129	1,324,242	2,089,874	-0.07	4.56
Cambodia	3,500	7,000	8,000	6.93	1.34
Indonesia	125,000	310,000	399,554	9.08	2.54
Lao PDR	720	3,843	3,700	16.75	-0.38
Malaysia	20,000	20,406	23,500	0.20	1.41
Myanmar	44,379	56,700	81,300	2.45	3.60
Philippines	573,150	269,270	392,720	-7.55	3.77
Thailand	494,080	520,153	930,000	0.51	5.81
Viet Nam	72,300	136,870	251,100	6.38	6.07
South Asia	3,827,407	4,053,494	2,683,397	0.57	-4.12
Afghanistan	4,000	3,000	2,000	-2.88	-4.05
Bangladesh	144,590	164,708	175,580	1.30	0.64
Bhutan	0	400	410		0.25
India	2,866,200	3,078,700	1,470,000	0.72	-7.39
Nepal	17,990	24,910	46,360	3.25	6.21
Pakistan	787,827	762,000	964,500	-0.33	2.36
Sri Lanka	6,800	19,776	24,547	10.68	2.16
Total Asia	5,837,463	6,367,752	8,627,226	0.87	3.04

Source: FAOSTAT Database. Agricultural Production Indexes. 22 April 1999. *Available: http://apps.fao.org*

Table A25: Output of Sugar Cane in Asia, 1977-1997

	Output (t)			Average Growth (percent per year)	
	1977	1987	1997	1978-1987	1988-1997
East Asia	**31,381,866**	**55,185,505**	**83,995,800**	**5.64**	**4.20**
China, People's Rep. of	29,053,866	52,811,505	82,565,800	5.98	4.47
Japan	2,328,000	2,374,000	1,430,000	0.20	-5.07
Southeast Asia	**81,086,060**	**76,687,595**	**118,049,470**	**-0.56**	**4.31**
Cambodia	195,000	164,000	187,500	-1.73	1.34
Indonesia	14,516,280	26,130,736	27,763,750	5.88	0.61
Lao PDR	17,840	112,853	95,000	18.45	-1.72
Malaysia	1,000,000	1,207,000	1,600,000	1.88	2.82
Myanmar	1,625,692	3,433,000	4,124,920	7.48	1.84
Philippines	34,820,000	17,600,000	27,000,000	-6.82	4.28
Thailand	26,094,448	24,449,936	45,850,100	-0.65	6.29
Viet Nam	2,816,800	3,590,070	11,428,200	2.43	11.58
South Asia	**189,719,149**	**224,259,298**	**329,831,243**	**1.67**	**3.86**
Afghanistan	64,000	60,000	38,000	-0.65	-4.57
Bangladesh	6,503,802	6,895,910	7,520,540	0.59	0.87
Bhutan	0	11,800	12,800		0.81
India	153,007,008	186,089,504	277,249,984	1.96	3.99
Nepal	311,379	616,580	1,629,300	6.83	9.72
Pakistan	29,522,960	29,925,808	41,998,400	0.14	3.39
Sri Lanka	310,000	659,696	1,382,219	7.55	7.40
Total Asia	**303,343,895**	**357,769,314**	**534,009,928**	**1.65**	**4.01**

Source: FAOSTAT Database. Agricultural Production Indexes. 22 April 1999. *Available: http://apps.fao.org*

Table A26: Yield of Sugar Cane in Asia, 1977-1997

	Yield (kg/ha)			Average Growth (percent per year)	
	1977	1987	1997	1978-1987	1988-1997
East Asia	**46,127**	**57,077**	**74,842**	**2.13**	**2.71**
China, People's Rep. of	46,127	57,077	74,842	2.13	2.71
Japan	71,411	68,023	63,556	-0.49	-0.68
Southeast Asia	**54,473**	**49,423**	**50,124**	**-0.97**	**0.14**
Cambodia	55,714	23,429	23,438	-8.66	0.00
Indonesia	116,130	84,293	69,487	-3.20	-1.93
Lao PDR	24,778	29,366	25,676	1.70	-1.34
Malaysia	50,000	59,149	68,085	1.68	1.41
Myanmar	36,632	60,547	50,737	5.02	-1.77
Philippines	60,752	65,362	68,751	0.73	0.51
Thailand	52,814	47,005	49,301	-1.17	0.48
Viet Nam	38,960	26,230	45,513	-3.96	5.51
South Asia	**35, 789**	**35,599**	**42,077**	**-0.05**	**1.67**
Afghanistan	16,000	20,000	19,000	2.23	-0.51
Bangladesh	44,981	41,868	42,833	-0.72	0.23
Bhutan	0	29,500	31,220		0.57
India	53,383	60,444	66,487	1.24	0.95
Nepal	17,308	24,752	35,145	3.58	3.51
Pakistan	37,474	39,273	43,544	0.47	1.03
Sri Lanka	45,588	33,358	56,309	-3.12	5.24
Asia	**51,965**	**56,185**	**61,898**	**0.78**	**0.97**

Source: FAOSTAT Database. Agricultural Production Indexes. 22 April 1999. *Available: http://apps.fao.org*

Table A27: Total Pork Production Per Capita in Selected Asian Economies

	Production (t)				Production Per Capita (t per 1,000)			Average Growth (percent per year)		
	1965	1975	1985	1995	1975	1985	1995	1966-1975	1976-1985	1986-1995
East Asia										
China, People's Rep. of	5,491,400	7,847,560	17,377,710	33,233,020	8.5	16.2	27.2	3.57	7.95	6.48
Japan	407,238	1,039,642	1,531,914	1,322,065	9.3	12.7	10.6	9.37	3.88	-1.47
Korea, Rep. of	76,272	98,848	433,750	799,000	2.8	10.6	17.8	2.59	14.79	6.11
Mongolia	500	600	2,200	600	0.4	1.2	0.2	1.82	12.99	-12.99
Taipei,China	na	na	na	na	na	na	na	na	na	na
Southeast Asia										
Cambodia	23,300	24,000	48,100	81,500	3.4	6.5	8.1	0.30	6.95	5.27
Indonesia	115,500	132,000	368,500	588,500	1.0	2.2	3.0	1.34	10.27	4.68
Lao PDR	13,400	9,000	17,800	25,038	3.0	5.0	5.1	-3.98	6.82	3.41
Malaysia	58,636	113,107	153,111	230,160	9.2	9.8	11.4	6.57	3.03	4.08
Myanmar	32,780	60,638	83,050	97,130	2.0	2.2	2.2	6.15	3.15	1.57
Philippines	290,000	323,000	396,942	969,862	7.5	7.3	14.3	1.08	2.06	8.93
Singapore	24,500	55,323	72,156	86,189	24.4	26.6	25.9	8.15	2.66	1.78
Thailand	150,000	175,000	377,500	301,090	4.2	7.4	5.2	1.54	7.69	-2.26
Viet Nam	275,000	247,000	560,668	1,007,000	5.1	9.4	13.6	-1.07	8.20	5.86
South Asia										
Afghanistan	na	na	na	na	na	na	na	na	na	na
Bangladesh	na	na	na	na	na	na	na	na	na	na
Bhutan	606	818	991	1,222	0.7	0.7	0.7	3.00	1.92	2.10
India	148,750	214,200	308,175	420,000	0.3	0.4	0.5	3.65	3.64	3.10
Maldives	na	na	na	na	na	na	na	na	na	na
Nepal	3,600	5,000	7,234	11,027	0.4	0.4	0.5	3.29	3.69	4.22
Pakistan	na	na	na	na	na	na	na	na	na	na
Sri Lanka	995	930	1,260	2,336	0.1	0.1	0.1	-0.68	3.04	6.17
Central Asia										
Kazakhstan	na	na	na	113,400	na	na	6.7	na	na	0
Kyrgyz Republic	na	na	na	17,900	na	na	4.0	na	na	0
Uzbekistan	na	na	na	16,000	na	na	0.7	na	na	0
Tajikistan	na	na	na	1,200	na	na	0.2	na	na	0
Turkmenistan	na	na	na	3,000	na	na	0.7	na	na	0
Total Asia	7,284,905	10,601,171	22,109,992	39,748,000				3.75	7.35	5.87

na = not available; 0 = zero or less than half of the unit measured.

Note: 1. Annual Growth Rate = ((Ln(value year begin) - Ln(value year end)) / number of years) x 100.

Source: FAOSTAT Database: Agricultural Production. Available: http://apps.fao.org

Table A28: Total Milk Production Per Capita in Selected Asian Economies

	Production (t)				Production Per Capita (t per 1,000)			Average Growth (percent per year)		
	1965	1975	1985	1995	1975	1985	1995	1966-1975	1976-1985	1986-1995
East Asia										
China, People's Rep. of	1,961,750	2,361,389	4,755,379	9,457,110	2.5	4.4	7.8	1.85	7.00	6.87
Japan	3,223,888	4,961,017	7,380,400	8,382,000	44.5	61.1	67.0	4.31	3.97	1.27
Korea, Rep. of	13,270	163,748	1,010,118	2,004,325	4.6	24.8	44.6	25.13	18.19	6.85
Mongolia	242,800	237,100	269,400	369,600	163.9	141.1	150.1	-0.24	1.28	3.16
Taipei,China	na	na	na	na	na	na	na	na	na	na
Southeast Asia										
Cambodia	15,045	17,850	16,320	19,125	2.5	2.2	1.9	1.71	-0.90	1.59
Indonesia	200,124	186,500	401,000	705,937	1.4	2.4	3.6	-0.71	7.66	5.66
Lao PDR	2,100	2,500	4,000	5,500	0.8	1.1	1.1	1.74	4.70	3.18
Malaysia	33,900	30,740	31,000	42,560	2.5	2.0	2.1	-0.98	0.08	3.17
Myanmar	161,554	265,154	650,672	555,586	8.7	17.3	12.3	4.95	8.98	-1.58
Philippines	26,341	31,179	33,000	34,000	0.7	0.6	0.5	1.69	0.57	0.30
Singapore	na	na	na	na	na	na	na	na	na	na
Thailand	2,300	8,200	57,895	270,000	0.2	1.1	4.6	12.71	19.54	15.40
Viet Nam	19,600	30,000	53,600	73,300	0.6	0.9	1.0	4.26	5.80	3.13
South Asia										
Afghanistan	737,530	848,000	670,100	549,600	55.1	46.2	28.0	1.40	-2.35	-1.98
Bangladesh	1,004,480	1,179,188	1,310,635	1,987,000	15.4	13.2	16.8	1.60	1.06	4.16
Bhutan	19,728	25,080	30,514	31,957	21.6	21.0	18.1	2.40	1.96	0.46
India	19,247,000	25,600,000	44,000,000	66,000,000	41.2	57.3	71.0	2.85	5.42	4.05
Maldives	na	na	na	na	na	na	na	na	na	na
Nepal	575,200	714,520	805,177	1,008,473	55.8	48.8	47.0	2.17	1.19	2.25
Pakistan	6,658,100	8,193,000	10,856,000	18,936,000	109.6	107.3	139.0	2.07	2.81	5.56
Sri Lanka	169,480	184,460	288,720	287,207	13.6	18.0	16.0	0.85	4.48	-0.05
Central Asia										
Kazakhstan	na	na	na	4,603,851	na	na	273.8	na	na	0
Kyrgyz Republic	na	na	na	864,000	na	na	193.7	na	na	0
Uzbekistan	na	na	na	3,780,000	na	na	166.1	na	na	0
Tajikistan	na	na	na	438,000	na	na	75.2	na	na	0
Turkmenistan	na	na	na	727,300	na	na	178.5	na	na	0
Total Asia	44,975,411	58,064,570	89,094,320	142,617,600				2.55	4.28	4.70

na = not available; 0 = zero or less than half of the unit measured.

Note: 1. Annual Growth Rate = ((Ln(value year begin) - Ln(value year end)) / number of years) x 100.

Source: FAOSTAT Database: Agricultural Production. Available: http://apps.fao.org

Table A29: Total Poultry Production Per Capita in Selected Asian Economies

	Production (t)				Production Per Capita (t per 1,000)			Average Growth (percent per year)		
	1965	1975	1985	1995	1975	1985	1995	1966-1975	1976-1985	1986-1995
East Asia										
China, People's Rep. of	525,700	814,350	6.72	6,002,313	0.9	1.3	4.9	4.38	5.36	14.62
Japan	204,340	739,873		1,267,000	6.6	11.2	10.1	12.87	6.04	-0.66
Korea, Rep. of	18,923	55,594	126,000	383,000	1.6	3.1	8.5	10.78	8.18	11.12
Mongolia	360	200	300	27	0.1	0.2	0.0	-5.88	4.05	-24.08
Taipei,China	na	na	na	na	na	na	na	na	na	na
Southeast Asia										
Cambodia	4,700	6,150	9,600	15,000	0.9	1.3	1.5	2.69	4.45	4.46
Indonesia	55,200	87,200	318,200	875,700	0.6	1.9	4.4	4.57	12.94	10.12
Lao PDR	6,740	3,864	5,160	9,064	1.3	1.4	1.9	-5.56	2.89	5.63
Malaysia	38,000	94,000	221,400	660,737	7.7	14.1	32.8	9.06	8.57	10.93
Myanmar	18,110	40,374	87,017	98,963	1.3	2.3	2.2	8.02	7.68	1.29
Philippines	80,000	131,627	178,398	399,557	3.1	3.3	5.9	4.98	3.04	8.06
Singapore	18,900	53,182	57,000	60,000	23.5	21.0	18.0	10.35	0.69	0.51
Thailand	117,000	253,000	393,000	780,000	6.1	7.7	13.4	7.71	4.40	6.85
Viet Nam	45,000	50,000	115,000	124,000	1.0	1.9	1.7	1.05	8.33	0.75
South Asia										
Afghanistan	6,800	10,400	12,800	13,680	0.7	0.9	0.7	4.25	2.08	0.66
Bangladesh	26,900	37,100	49,011	89,481	0.5	0.5	0.8	3.21	2.78	6.02
Bhutan	61	111	238	329	0.1	0.2	0.2	5.99	7.63	3.24
India	71,001	88,020	161,100	478,800	0.1	0.2	0.5	2.15	6.04	10.89
Maldives	na	na	na	na	na	na	na	na	na	na
Nepal	3,800	4,600	5,299	9,396	0.4	0.3	0.4	1.91	1.41	5.73
Pakistan	11,500	22,000	102,240	308,000	0.3	1.0	2.3	6.49	15.36	11.03
Sri Lanka	11,000	16,000	18,600	54,000	1.2	1.2	3.0	3.75	1.51	10.66
Central Asia										
Kazakhstan	na	na	na	53,000	na	na	3.2	na	na	0
Kyrgyz Republic	na	na	na	2,700	na	na	0.6	na	na	0
Uzbekistan	na	na	na	16,000	na	na	0.7	na	na	0
Tajikistan	na	na	na	1,700	na	na	0.3	na	na	0
Turkmenistan	na	na	na	4,000	na	na	1.0	na	na	0
Total Asia	1,512,179	3,078,570	6,027,808	13,951,590				7.11	6.72	8.39

na = not available; 0 = zero or less than half of the unit measured.

Note: 1. Annual Growth Rate = ((Ln(value year begin) - Ln(value year end) / number of years) x 100.

Source: FAOSTAT Database: Agricultural Production. *Available: http://apps.fao.org*

Table A30: Total Egg Production Per Capita in Selected Asian Economies

	Production (t)				Production Per Capita (t per 1,000)			Average Growth (percent per year)		
	1965	1975	1985	1995	1975	1985	1995	1966-1975	1976-1985	1986-1995
East Asia										
China, People's Rep. of	1,653,809	2,307,742	5,540,540	17,082,970	2.5	4.5	14.0	3.33	8.76	11.26
Japan	1,330,000	1,807,000	2,152,356	2,550,590	16.2	17.2	20.4	3.06	1.75	1.70
Korea, Rep. of	65,273	176,160	316,680	456,284	5.0	7.1	10.2	9.93	5.86	3.65
Mongolia	290	420	1,330	180	0.3	0.5	0.1	3.70	11.53	-20.00
Taipei,China	na	na	na	na	na	na	na	na	na	na
Southeast Asia										
Cambodia	5,000	8,100	9,550	13,250	1.1	1.3	1.3	4.82	1.65	3.27
Indonesia	60,000	111,000	369,300	732,000	0.8	2.2	3.7	6.15	12.02	6.84
Lao PDR	1,283	3,274	5,306	6,252	1.1	1.5	1.3	9.37	4.83	1.64
Malaysia	39,600	105,700	186,300	368,069	8.6	11.9	18.3	9.82	5.67	6.81
Myanmar	17,732	26,641	58,803	54,351	0.9	1.6	1.2	4.07	7.92	-0.79
Philippines	92,611	194,628	245,120	366,000	4.5	4.5	5.4	7.43	2.31	4.01
Singapore	14,150	24,310	21,100	19,260	10.7	7.8	5.8	5.41	-1.42	-0.91
Thailand	314,000	382,000	502,500	705,000	9.2	9.8	12.1	1.96	2.74	3.39
Viet Nam	107,000	101,000	137,000	216,000	2.1	2.3	2.9	-0.58	3.05	4.55
South Asia										
Afghanistan	12,800	13,370	14,200	14,700	0.9	1.0	0.7	0.44	0.60	0.35
Bangladesh	38,600	56,500	68,400	110,000	0.7	0.7	0.9	3.81	1.91	4.75
Bhutan	110	140	240	370	0.1	0.2	0.2	2.41	5.39	4.33
India	207,000	460,000	887,000	1,500,000	0.7	1.2	1.6	7.99	6.57	5.25
Maldives	na	na	na	na	na	na	na	na	na	na
Nepal	9,500	13,400	12,705	19,600	1.0	0.8	0.9	3.44	-0.53	4.34
Pakistan	9,690	42,100	169,519	285,100	0.6	1.7	2.1	14.69	13.93	5.20
Sri Lanka	16,290	16,798	35,038	48,928	1.2	2.2	2.7	0.31	7.35	3.34
Central Asia										
Kazakhstan	na	na	na	103,300	na	na	6.1	na	na	0
Kyrgyz Republic	na	na	na	8,278	na	na	1.9	na	na	0
Uzbekistan	na	na	na	69,400	na	na	3.0	na	na	0
Tajikistan	na	na	na	3,565	na	na	0.6	na	na	0
Turkmenistan	na	na	na	69,400	na	na	17.0	na	na	0
Total Asia	4,322,566	6,446,128	11,906,807	26,525,940				4.00	6.14	8.01

na = not available; 0 = zero or less than half of the unit measured.

Note: 1. Annual Growth Rate = ((Ln(value year begin) - Ln(value year end)) / number of years) x 100.

Source: FAOSTAT Database: Agricultural Production. *Available: http://apps.fao.org*

Table A31: Asia's Top Seven Fish Producers, ranked by 1987–1996 production

	Production (t)			Share in Total Production (%)			Average Annual Growth (%)		Average Production (t)
	1976	1986	1996	1976	1986	1996	1977-1986	1987-1996	1987-1996
Total Production (fish, shellfish)									
China, People's Rep. of	4,632,744	8,721,497	31,936,876	16.0	20.9	47.6	6.3	13.0	18,170,016
Japan	9,996,833	11,976,493	6,793,444	34.5	28.7	10.1	1.8	-5.7	9,221,498
India	2,177,096	2,937,470	5,260,420	7.5	7.1	7.8	3.0	5.8	4,121,131
Indonesia	1,478,926	2,456,972	4,401,940	5.1	5.9	6.6	5.1	5.8	3,431,281
Thailand	1,660,386	2,540,835	3,647,900	5.7	6.1	5.4	4.2	3.6	3,145,878
Korea, Rep. of	2,117,755	3,103,486	2,771,772	7.3	7.4	4.1	3.8	-1.1	2,731,268
Philippines	1,393,420	1,916,347	2,133,063	4.8	4.6	3.2	3.2	1.1	2,170,404
Top Seven	23,457,160	33,653,100	56,945,415	81.0	80.7	84.9	3.6	5.3	42,991,475
Other	5,505,886	8,038,708	10,167,385	19.0	19.3	15.1	3.8	2.3	9,592,962
Total Asia	28,963,046	41,691,808	67,112,800	100.0	100.0	100.0	3.6	4.8	52,584,437
Marine Capture (fish, shellfish)									
Japan	na	11,178,058	5,870,585	na	35.6	15.1	na	-6.4	8,317,974
China, People's Rep. of	na	4,165,452	12,459,446	na	13.3	32.0	na	10.9	7,570,535
Thailand	na	2,305,378	2,934,600	na	7.3	7.5	na	2.4	2,630,504
Indonesia	na	1,849,866	3,385,440	na	5.9	8.7	na	6.0	2,589,454
India	na	1,716,944	2,840,919	na	5.5	7.3	na	5.0	2,333,282
Korea, Rep. of	na	2,623,514	2,405,692	na	8.3	6.2	na	-0.9	2,320,133
Korea, Dem. People's Rep. of	na	1,559,239	1,622,000	na	5.0	4.2	na	0.4	1,596,160

Table A31 (cont.)

Table A31: Asia's Top Seven Fish Producers, ranked by 1987–1996 production

	Production (t)			Share in Total Production (%)			Average Annual Growth (%)		Average Production (t)
	1976	1986	1996	1976	1986	1996	1977-1986	1987-1996	1987-1996
Marine Capture (fish, shellfish)									
Top Seven	na	25,398,451	31,518,682	na	80.8	81.0	na	2.2	27,358,042
Other	na	6,026,029	7,380,329	na	19.2	19.0	na	2.0	7,227,132
Total Asia	na	31,424,480	38,899,011	na	100.0	100.0	na	2.1	34,585,174
Aquaculture (fish, shellfish)									
China, People's Rep. of	na	3,962,273	17,714,570	na	54.7	75.5	na	15.0	9,505,535
India	na	686,260	1,768,422	na	9.5	7.5	na	9.5	1,271,712
Japan	na	692,762	829,354	na	9.6	3.5	na	1.8	802,003
Indonesia	na	333,092	672,130	na	4.6	2.9	na	7.0	530,400
Korea, Rep. of	na	428,212	358,003	na	5.9	1.5	na	-1.8	389,475
Philippines	na	302,055	342,678	na	4.2	1.5	na	1.3	369,703
Thailand	na	134,057	509,656	na	1.9	2.2	na	13.4	367,734
Top Seven	na	6,538,711	22,194,813	na	90.3	94.6	na	12.2	13,236,562
Other	na	705,157	1,270,254	na	9.7	5.4	na	5.9	1,024,707
Total Asia	na	7,243,868	23,465,067	na	100.0	100.0	na	11.8	14,261,269

Note: Average growths (1977-1986 and 1987-1996) were calculated from time-series data presented in FAO (1998b, 1998c).

Table A32: Marine Landings of Miscellaneous Fishes (ISSCAAP 39) in Selected Economies and Fishing Areas, 1986–1996

	Fishing Area	Production (t) 1986	Production (t) 1996	Average Annual Growth (%) (1987–1996)	% Share 1996	% of Total 1986	% of Total 1996
China, People's Rep. of	Pacific, NW	1,471,029	4,115,836	10.29	38.89	35.31	33.03
Korea, Dem. People's Rep. of	Pacific, NW	1,559,000	1,621,800	0.39	15.33	99.98	99.99
Thailand	Pacific, WC	919,482	901,751	-0.19	8.52		
Thailand	Indian, E	207,913	349,750	5.20	3.30		
Thailand, total		1,127,395	1,251,501	1.04	11.83	48.90	42.65
India	Indian, W	134,066	561,596	14.32	5.31		
India	Indian, E	58,051	166,598	10.54	1.57		
India, total		192,117	728,194	13.32	6.88	11.19	25.63
Japan	Pacific, NW	577,020	287,326	-6.97	2.72		
Japan, other		638,952	622,716	-0.26	5.88		
Japan, total		598,230	292,560	-7.15	2.76	5.35	4.98
Myanmar	Indian, E	528,158	613,530	1.50	5.80	98.69	96.54
Viet Nam	Pacific, WC	448,177	412,000	-0.84	3.89	79.88	60.23
Indonesia	Pacific, WC	215,861	337,000	4.45	3.18		
Indonesia	Indian, E	43,137	56,000	2.61	0.53		
Indonesia, total		258,998	393,000	4.17	3.71	14.00	11.61
Malaysia	Indian, E	163,341	177,897	0.85	1.68		
Malaysia	Pacific, WC	46,516	174,941	13.25	1.65		
Malaysia, total		209,857	352,838	5.20	3.33	27.23	31.32
Korea, Rep. of	Pacific, NW	74,551	148,719	6.91	1.41		
Korea, Rep. of	Pacific, WC	335	11,628	35.47	0.11		
Korea, Rep. of, other		20,518	18,208	-1.19	0.17		
Korea, Rep. of, total		95,404	178,555	6.27	1.69	3.64	7.42
Subtotal		6,488,365	9,959,814	4.29	94.12	23.80	29.32
Other countries		21,210	5,234	-13.99	0.05	0.51	0.11
Total Asia		**7,127,317**	**10,582,530**	**3.95**	**100.00**	**22.68**	**27.21**

Legend: E: Eastern, W: Western, NW: Northwest, WC: Western Central

Notes: Large percentages of miscellaneous fish catches of some countries may be due to inadequate data reporting.

Average growth (1987-1996) was calculated using time-series data in FAO (1998c).

Table A33: Marine Capture Fisheries Production and Growth in Fishing Areas surrounding Asia

FAO Fishing Area	Production (t) 1986	Production (t) 1996	Average 1987–1996 (%)	Av. Growth 1987–1996 (%/year)
61 - Pacific, Northwest	18,080,841	21,753,037	54.70	1.85
71 - Pacific, Western Central	6,135,980	8,516,979	22.16	3.28
57 - Indian Ocean, Eastern	2,503,465	3,748,591	9.05	4.04
51 - Indian Ocean, Western	2,286,694	3,499,009	8.77	4.25
37 - Mediterranean & Black Sea	548,995	490,710	1.47	-1.12
Major Fishing Areas - Aggregate	29,555,975	38,008,326	96.14	2.52
Other	1,868,505	890,685	3.86	-7.41
Asia, Total Marine Capture	**31,424,480**	**38,899,011**	**100**	**2.13**
World, Total Marine Capture	**78,568,379**	**87,072,588**	—	**1.00**

Notes: Mediterranean and Black Sea are the local fishing grounds of some West Asian countries.
Average growth (1987-1996) was calculated using time-series data in FAO (1998c).

Table A34: Marine Capture Fisheries Production in Asia by Species Group

Species Group*	Production 1986 (t)	Production 1996 (t)	Percentage of Total 1986	Percentage of Total 1996	Av. Growth (1987–1996) (%/year)
Mainly Pelagic (31, 32, 33)	13,926,617	15,003,218	44.32	38.57	0.74
Mainly Demersal (34, 35, 36, 37, 38)	5,413,847	4,934,184	17.23	12.68	-0.93
Misc. Marine Fishes (39)	7,127,317	10,582,530	22.68	27.21	3.95
Total Marine Fishes	26,467,781	30,519,932	84.23	78.46	1.42
Crustaceans, Molluscs, Cephalopods, etc.	4,601,829	7,706,538	14.64	19.81	5.16
Salmon, Trout, Shads, Misc. Diadromous (23, 24, 25)	354,870	672,541	1.13	1.73	6.39
Asia, Total Marine Capture	**31,424,480**	**38,899,011**	**100.00**	**100.00**	**2.13**

Notes: Species groups are ISSCAAP Species Groups. The designation as mainly pelagic or demersal is indicative only because some species, e.g., sharks, berbels, etc. (No. 38) can be both pelagic and demersal.
Average annual growth rates (1987-1996) were calculated using time-series data in FAO (1998c).

Annex B

SUMMARY OF SUPPORTING RECOMMENDATIONS

The following Table provides a summary of recommendations and suggestions described in the text but not elaborated as strategies in Chapter VI.

Sector	Recommendations
Rice/Annual Crops	• Undertake R&D to explore possibilities and means (including biotechnology) to decrease effectively the production costs (per kg) of major crops, i.e. rice, wheat, maize, and soybean, in favorable environments. • Undertake RD&E for integrated crop and soil management for less favorable environments, with special emphasis on the ability of the RD&E system to respond to local and emerging problems and opportunities. • Establish breeding programs to increase yield and lower cost of production for rice, soybean, and pulses in less favorable environments (tolerance to drought, phosphorus deficiency, and tolerance to soil acidity).

Sector	Recommendations
Rice/Annual Crops (cont.)	• Identify factors and conditions responsible for yield gaps for rice and the feasibility of closing the gaps. • Intensify efforts to raise the yield potential of rice (including use of biotechnology).
Oil Palm / Fire Haze	• Locate land for palm oil plantations outside natural forests (Indonesia). • Recognize land rights of local communities. • Revise promotion incentives and conditions for large-scale plantations. • Revise export policies for forest products. • Strengthen fire protection capacity.
Rice/ Fire Haze	• Ban rice production in peat swamps (Indonesia).
Water	• Undertake institutional reform for integrated planning and management. • Adopt a basin approach for water resources planning and management. • Adopt participatory planning and management of water resources. • Eliminate open-access conditions through the use of market-based instruments or social and legal means. • Improve irrigation efficiency. • Rehabilitate existing irrigation infrastructure.

Sector	Recommendations
Water (cont.)	• Invest in research, monitoring, EIA, and SIA. • Enforce the polluter-pays principle. • Include improvement of capacity for economic analysis in water agencies in water projects. • Improve the capacity for EIA and SIA for water projects. • Improve supply-and-demand management.
Livestock	• Provide incentives to relocate farms and cooperatives through the use of economic instruments. • Provide incentives for the use of low-cost biogas technology. • Provide training to farmers and cooperative members. • Abolish subsidies on fossil fuels. • Enforce pollution control regulations. • Include consideration of the sustainability of breeding systems.
Fisheries	• Review and regulate the capacity of fishing fleets in relation to sustainable yields of fishery resources and where necessary reduce fleet size. • Adopt policies, application of measures, and development of techniques to reduce bycatch, fish discards, and postharvest losses. • Develop ecologically sound aquaculture as an important contributor to overall food security.

Sector	Recommendations
Fisheries (cont.)	• Promote culture fisheries for low-income groups. • Strengthen fisheries research and increase cooperation among research institutions. • Increase efforts to estimate stock sizes as well as the quantity of fish and other organisms caught incidentally and discarded during fishing operations. • Strengthen research programs aiming to stimulate environmentally sound aquaculture and stocking, with special emphasis on: (1) the impact on the environment and biodiversity; (2) the application of biotechnology; and (3) the health of cultured stocks. • Improve sanitation in landing facilities, especially for small-scale fisheries.
Agroforestry	• Promote exchange of agroforestry experience. • Revise policies and incentives for agroforestry. • Improve access to government R&D and farmer-generated knowledge. • Improve market information for agroforestry.
Forest Policy	• Undertake institutional reform to reflect total value of forests, not only the value of timber. • Provide capacity building in the area of conservation.

Sector	Recommendations
Forest Policy (cont.)	• Review pricing of concessions. • Review nonforestry (e.g. export) policies that encourage expansion of agricultural activities in forests.
Protected Areas	• Implement participatory management, sharing responsibility with related agencies and local communities. • Develop financing options, such as entrance fees, debt for nature swaps, conservation funds, and nondevelopment rights. • Ensure proper valuation in projects on impact on protected areas.
Agricultural Extension	• Explore the possibility and necessary means, including the involvement of NGOs and the private sector, to provide effective linkages between problem (local and emerging) identification and problem solving. • Explore the feasibility of establishing services within the fertilizer industry for soil and plant analysis and recommendations for efficient fertilizer use. Identify needed inputs (technical support), incentives (financial and tax advantages), and mechanisms necessary for quality control and minimization of offsite impact in fertilizer use.

Sector	Recommendations
Agricultural Extension (cont.)	• Provide avenues to facilitate farmer-to-farmer technology transfer. • Use distance media to reach special groups and women.
Land Policy	• Provide better land security. • Use tax instruments to release unused land for cultivation. • Recognize community management of rangelands.

Author Index

Subject Index